A Practical Guide to Logistics

A Practical Guide to Logistics

An introduction to transport, warehousing, trade and distribution

Jerry Rudd

First published in Great Britain and the United States in 2019 by Kogan Page Limited

2nd Floor, 45 Gee Street
London
EC1V 3RS
United Kingdom

122 W 27th St, 10th Floor
New York, NY 10001
USA

4737/23 Ansari Road
Daryaganj
New Delhi 110002
India

www.koganpage.com

ISBNs

Hardback 9780749498818
Paperback 9780749486310
Ebook 9780749486327

British Library Cataloguing-in-Publication Data

A CIP record for this book is available from the British Library.

Library of Congress Cataloging-in-Publication Data

A CIP record for this book is available from the Library of Congress.

Typeset by Integra Software Services, Pondicherry
Print production managed by Jellyfish
Printed and bound by CPI Group (UK) Ltd, Croydon CR0 4YY

CONTENTS

FIGURES

TABLES

ACKNOWLEDGEMENTS

When I first retired from a highly enjoyable career in logistics, the idea of writing a book was given little more than a fleeting thought. However, one of my longest-standing business associates, Ruth Waring FCILT (now Director at BigChange Advisory Plus), provided the spark that led to me developing the idea for this book, and turning that fleeting thought into a reality. I have genuinely enjoyed writing the book, so I thank her sincerely. I would also like to thank Julia Swales, Head of Publishing at Kogan Page, for accepting my proposal, and my Development Editor, Rajveer Ro'isin Singh, for her assistance during the preparation of the manuscript.

A number of people have been kind enough to meet with me or have a telephone conversation to share their expertise in certain areas, and to update me on recent developments. They include Kevin Sarpong of Linde (regarding materials handling equipment); Jon Hugill of Hallett Silbermann (abnormal loads); Lindsay Durham of Freightliner (railfreight); Tim Harding of Gefco (customs regulations); Nick Betteley of Stakapal (racking systems); Joe Daniel of JZD Solutions (tankers); Matt Garland of Palletline (pallet networks); Malcolm Latarche FICS, Journalist (ship charter); Guy Bates of Network Rail (railfreight subsidies); John Tivey of Marlow, Gardner and Cooke (insurance) and Pierre van der Stichele of Chapman Freeborn (air charter). Others have been kind enough to discuss the products or services offered by their own respective organizations, including Jane Geary of Maxoptra; Olivia Bannister of Flogas; Bob Morris of Isuzu; Steven McLachlen of Scania; Jon Millatt of Scotline; James Young of Transmec and Alice Perroni of Techsert.

It would not have been able to provide as many illustrations without the help of a great many people in providing suitable pictures, and/or granting permission to publish them. They include Nicki Whittaker of Prologis; Eddie Ping of Abbott Wade; John Stears of Stears Haulage; Markus Fischbacher of SEP Logistics AG; Nick Betteley of Stakapal; Kate Colclough of Gonvarri; Neil Emmott of Avanta; David Sheppard and Barry Wilkinson of Storax; Elizabeth Phillipson of Herchenbach; Terry Siddle of Pets at Home; Michaela Jupp of Linde; Robin Kidner of Safety Lifting Gear; Amanda Cavan and Philip Shaw of The Ramp People; Mark James of Thorworld; Andrea Munt of Loading Systems; Darrell Hopwood of Armo; Olivia Bannister of Flogas; Brian Owens of Albion Handling; Karen and Matthew Nickson of Axiom; Victoria Vincent of B and B Attachments; Cameron Moir of Sotrex; Jonathan Edgar and Rhiannon Prothero of SAP; Mari Still of Datalogic; Mike Alibone of SSI Schäfer; Amy Southall of Crown Lift Trucks; Valeria Mills and Jackie Woodward of Macfarlane Packaging; Danielle Cunningham and Paddy McCartney of Robopac; Eloise Edgar of Rajapack; Richard Loynds of Yolli; Ellie Filcock of

Davpack; Colin Woodhouse of Barnes and Woodhouse; Richard Jones of Premier Pallet Systems; Janet Lowe of Lowe RP; Alex Hollamby of PPS Equipment; Rob Barton of Barcode Warehouse; Charlotte Ward of Ford Motor Company; James Emberson of Isuzu; Martin Radford of LC Vehicle Hire; Julie Daintith of Eddie Stobart; Nigel Hanwell of Volvo; Ross Hyslop of Annandale; Carl Wilson of CW Marketing on behalf of Avon Material Supplies; Peter Denby and John Gallagher of Denby Transport; Matt Garland and Lorna Cox of Palletline; James Dennison of Dennison Trailers; Michael Fox of Abbey Logistics; Jon Hugill of Hallett Silbermann; Adam Kirk of Bredneck; Wendy Vaughan of Checkpoint; Mike Garside of UK Drug Testing; Jemma James of Trutac; Jane Geary of Maxoptra; John Keefe of Eurotunnel; Tanja Kaufmann of Ralpin; Jane Dry and Lynn Crump of Freightliner; Dick Heijstek of Mercarius; Joao Pinheiro and Gary Walker of CLdN; Louise Mason Blything of Universeal; Michael Bergh of Unifeeder, and Dawid Bujewski who took the photograph; Derek Cawston of Goldstar; Jon Millatt of Scotline; Piet Vandenkerkhove of the Port of Zeebrugge; and Andy James of Chapman Freeborn.

I thank Steve Hobson for granting permission to reproduce information from the Motor Transport Cost Tables.

I would also like to thank Paula Devine of the Bedfordshire Chamber of Commerce for allowing me to visit their offices in Luton and access their reference materials, and the staff of the Social Sciences Reading Room at the British Library.

I also thank my father, Derek Rudd, for initially stimulating my interest in transport and logistics, by taking me as a small child to watch the trains at Peterborough Station and ships being loaded in Great Yarmouth.

Finally, I must thank my wife, Diane Rudd, for her unfailing and invaluable support throughout my career, and during the writing of this book. She has been kind enough to proofread the manuscript, and to tolerate my renewed enthusiasm for all things logistical, including even my comments about the type of racking used in an episode of *Doctor Who*.

Introduction 01

Leaders win through logistics. Vision, sure. Strategy, yes. But when you go to war, you need to have both toilet paper and bullets at the right place at the right time. In other words, you must win through superior logistics.

TOM PETERS

Tom Peters, author of *In Search of Excellence* and one of the world's leading business writers, understands the vital importance of logistics. This is equally true in business as in war. A robust logistics system is an essential contribution to the success of any enterprise, without which it cannot hope to compete in the modern marketplace.

So, what is 'logistics'?

A down-to-earth definition was the tongue-in-cheek comment made by the managing director of a national retailer to an audience of senior logistics managers some years ago – 'All you lot do is run sheds and lorries.' Some in the audience might have essentially agreed with Tom Peters, and said it means 'Ensuring that the right quantities of the right items are in the right place at the right time at the right cost.' Both contain a large element of truth. Whilst the discipline of logistics is broader than the physical activities of running warehousing and transport, and not all transport movements are made by trucks, these elements represent its core. It is also fairly common for organizations to restrict their definition of logistics to just these core elements, placing them under the remit of their logistics function (maybe using suppliers), whilst handling the remainder elsewhere in-house, perhaps unwittingly creating a situation in which their subcontractors do indeed only run sheds and lorries.

One of the most important points to remember is that the principles of logistics remain constant regardless of the commodity or nature of the business. Items are sourced; they are transported to a storage location; they are worked upon, sometimes involving processes that fall outside the remit of logistics; they are delivered to a customer or end user; and all necessary information flows are correctly managed.

Whilst the nature of these stages may vary enormously, that is logistics. A wholesaler of electrical goods may import completed units from the Far East, store them in a warehouse for weeks or months, make little change other than adding a label, then deliver to retailers on a conventional curtain-sided articulated lorry (artic). A car manufacturer may bring in thousands of parts for each vehicle, use many hours of labour to assemble the car, store it in an outdoor compound and deliver to a dealer using a car transporter. A wheat farmer may buy seed, nurture and harvest the crops, store the corn in a grain silo and deliver to a flour mill using a bulk tipper truck. In all cases the movements must be properly managed and any necessary legal requirements duly complied with. The equipment used may vary, and some special rules may apply (for example, vehicles carrying grain must not previously have been used to carry certain potential contaminants such as glass), but the basic principles remain unchanged.

The purpose of this book is to lay out the various options for completing each stage of the process and help practitioners to select the best option for their particular business. I will be discussing both physical infrastructure and equipment, and operational processes. Other important topics such as packaging and customer requirements; import, export and customs documentation; trading terms; and inventory control will also be covered. There are many books on logistics already in print, some of them giving excellent in-depth analysis of particular aspects suitable for those preparing masters' theses. It is my aim to cover the issues at a more practical level, and I hope this book will be of interest to experienced and inexperienced practitioners alike.

You can download supporting material for this book at **koganpage.com/PGLog**.

First steps – strategic decisions 02

Should we operate our own logistics or contract out?

One of the longest-running debates in the industry is whether it is better to operate an in-house logistics network, with directly employed people, based on company-owned or rented property, using company-owned vehicles and other assets; or whether to outsource to one or more third party logistics suppliers (3PLs), using their people and assets. If using 3PLs, there are further choices to be made regarding dedicated or shared user services, and whether charging should be on an open book or closed book basis – these will be discussed later.

There is in my opinion no right or wrong answer, and in practice few companies operate a wholly in-house or wholly outsourced network – they use a combination of the two. The major UK retail group Sainsbury and Argos, for example, operate 34 distribution centres in the UK (as at 2018) – some of these are managed in-house, and others by 3PL suppliers such as Wincanton, NFT and DHL. At the other end of the size spectrum, a small business may decide to operate its own stores with a single operative, and leave its deliveries to one of the major parcels carriers.

If you say the words 'third party' to the average person, most will instinctively think about third-party car insurance. The first party is the person insuring the car, the second is the insurance company. The policy only comes into play when a third party wants to make a claim, for example the driver of another car involved in an accident. Third party has a similar meaning in logistics. A company may deliver its product to its customer on its own vehicle, or the customer may collect, but in many cases they will use the services of a 3PL company to provide the transport and/or warehousing.

The main arguments in favour of a 3PL are:

- Core expertise and flexibility. Small and medium size enterprises (SMEs) in particular have often grown due to the energy and enthusiasm of one or a few founders, who may well know all there is to know about their particular product, be it an innovative electronic gadget, or a delicious food, but logistics is not their core expertise. (The same is amazingly true of some larger and better-known organizations also!) By employing the services of a 3PL they can leverage the skills of people whose core expertise is logistics, and make dramatic improvements in both cost and quality.

- Reduced capital expenditure and liabilities. In an expanding business, cash flow may be tight. Capital for investment in warehouses, forklifts and vehicles, or even for their lease, may not be available, but a 3PL may well have the resources already at hand.
- Flexibility. Some would say that this is the biggest advantage of using a 3PL. The business may need to store an average of 1,000 pallets at present, but have expansion plans that will necessitate 2,000 pallets to be stored in three years' time. For an in-house operation the options are to set up a warehouse for 1,000 pallets, which will create a problem when the business grows; or establish a warehouse for 2,000 pallets, with wasted costs during the period the business is growing. Added to this are potential problems if the business fails to grow as expected, perhaps only reaching 1,500 pallets, or it may become a victim of its own better-than-expected success, with a 3,000-pallet requirement. A 3PL may well be able to accommodate this variation, and take the risk of greater or smaller than expected growth.
- Management time. In any business the unexpected happens, and this is true of logistics as much as any other aspect. Vehicles break down, drivers take time off due to sickness and leave for other jobs and goods are damaged in transit. If using a 3PL, the problem of finding a replacement vehicle or driver rests with that 3PL, and a claim for the cost of the damaged goods can be made against them.
- Finally, one must not forget operator licensing. More will be said on this subject later, but in short, any business that operates one or more vehicles bigger than 3.5 tonnes (the size of a long wheelbase transit van) must by law obtain an operator's licence before doing so. It is not legally possible to simply buy or rent a truck and start operating it.

Against all of this, the main arguments in favour of an in-house operation are:

- Control and peace of mind that the business will never be anything other than the 100 per cent priority of the operation. There is sometimes a perception, which may or may not be justified, that if a 3PL has more than one customer, it will favour the other one. If there is a shortage of capacity, for example due to a vehicle break down, it may be necessary to decide which of the 3PLs customers receives an on-time delivery and which will have to wait whilst contingency actions are taken (such as hire of a replacement vehicle) and receive a later delivery. With an in-house operation, if difficult decisions need to be made about what should be the priority, there are no competing external factors to consider.

- Motivation of the workforce. Many companies feel that they engender a team spirit amongst their workforce such that they all share a genuine commitment to the success of the business. They believe that they can motivate their own staff better than any third party, and have faith in their own ability to avoid or resolve problems such as employee absence rather than relying upon an outside entity. They may also believe that they can impart better product knowledge to their own people. Specialized personnel, such as a logistics manager and forklift drivers, can be recruited if necessary, and imbued with enthusiasm and the company ethos. If a company is genuinely able to do so, then this is a powerful argument in favour of an in-house solution.
- Psychological factors, including being able to be close to one's products. Many SMEs have a warehouse close or even adjacent to their offices so that they can access their goods at will. This may be attractive to the sales team if a potential customer is visiting, or if there is a technical query that is most easily resolved by physically examining the stock.
- Flexibility of 3PLs may in practice be limited. Most shared-user warehouses have lots of free space in February. However, this will not be the case in November. If your business happens to have a peak demand leading up to Christmas at the same time as most others then at the very least careful planning will be required. I know of one case in which a fashion retailer (with an autumn/winter peak) and a garden suppliers manufacturer (with a spring/summer peak) shared a third party warehouse. Both were household names. Space, forklifts and labour could all be shared between the two, which were an almost perfect match. However, such matches are in practice a rarity.
- Cost. I will talk about specific costs later, but will make some general observations here. 3PLs must make a profit in order to survive, and they will have overheads that must be paid for, such as personnel and accounts departments, and the customer must in practice pay a part of the board's salaries. Against this, an in-house operation may have similar costs. For example, if there are just a handful of additional people involved then the payroll department will undoubtedly be able to absorb the workload. However, if there are hundreds of personnel then additional payroll staff will be required and this will need to be taken into account in any cost calculation.
- Opportunity cost, which is also important. There may be space on the premises to accommodate a warehouse, but could that space be used in a different way to generate greater profits? Last but not least, assets are expensive. Whilst it is of course possible to lease or rent equipment, if buying a new 44 tonne artic unit and trailer in 2018 there will be little if any change from £100,000.

In conclusion, there is no right answer that applies to every case. In practice, most companies operate a mixed system, even the largest. A business may decide to run its own warehouse but contract out transport, or may have its own truck fleet for deliveries in most of England but subcontract Scotland, Wales and the South West. Some even provide the building and physical infrastructure for a warehouse but use a 3PL to provide management and labour. Similarly, domestic road movements may be performed in-house but international movements contracted to a 3PL. There are very few companies who operate air- and seafreight in-house.

Every decision needs to be considered on its individual merits, and I hope this discussion will have helped to lay out the factors to be considered if you need to make such a choice.

Dedicated or shared user?

The first point to make is that the choice between a dedicated or shared user is not the same as the decision whether to operate in-house or contract out to a 3PL. Many vehicles seen on the roads that bear a company's livery and have a driver wearing their uniform are in fact operated by 3PLs.

A dedicated warehouse or vehicle will be used only for the benefit of a single business. Such a warehouse will usually be in a building of its own, with a workforce working only on that operation, although it is also possible to occupy only a dedicated part of a building. An example of this was a 3PL warehouse in Coventry that had over 120,000 square feet (11,000 square metres) of shared user space accommodating goods from stationery to car parts. However, approximately 1,000 square feet (90 square metres) in a separate room was dedicated to the storage and handling of surgical implant artificial lenses used in cataract surgery, and associated surgical tools and pharmaceuticals. This operation had very specialized requirements, from hygiene to the tightest possible batch control and tracking, with a high level of training and expertise required for its staff. It was thus a totally dedicated operation even though it was on a shared user 3PL site.

Similarly, dedicated vehicles will carry goods only for a particular business, whether it is a single van delivering around town for the local flower shop or a fleet of hundreds of tankers delivering petrol and diesel for a major oil company.

Dedicated solutions tend to be the best option where requirements are large, regular, self-contained and/or specialized; shared user solutions tend to come to the fore where flexibility is the key – provided, of course, that it is viable.

Extreme examples of the former can be found at Daventry, where Sainsbury's and Tesco have nearby sites of 1 million and 750,000 square feet (93,000 and 70,000 square metres) respectively (Figure 2.1). It is self-evidently not viable to accommodate such large facilities in a shared user environment. Similarly, in 2017 John Lewis opened a new 638,000 square foot (59,000 square metre) distribution centre called Magna Park 3 in Milton Keynes, which I have had the pleasure of visiting. It handles electricals, furniture, bedding and leisure goods, amongst other items. The main year-round operation is highly automated, and very impressive to view. However, a large part of the facility is also laid out with the necessary equipment for a manual operation, which is intended to be brought into use in the busier times of the year, mainly the run-up to Christmas. It appears to me that John Lewis have decided that the cost of keeping part of the property and some equipment idle for part of the year is justified because it would be extremely difficult to find a suitable off-site alternative that could guarantee the level of service with which John Lewis would wish to be associated.

To consider vehicle operations, regular return trips and shorter distance routes are often best made on a dedicated basis. Major car manufacturers make nightly parts deliveries to all their dealers, and whilst the volumes tend to be higher in winter than summer, the routes are almost unchanging and ideally suited to a dedicated fleet of vans. Similarly, if an SME is making regular deliveries to the same regular customers then a dedicated fleet is again likely to be cost-effective. If the work is to be regular, and year-round business for a vehicle can be guaranteed, then a favourable rate should be obtainable from a 3PL.

Figure 2.1 Sainsbury's distribution centre at Daventry

SOURCE Reproduced by kind permission of Prologis UK

For longer distance work, a dedicated vehicle is only likely to be viable on a round trip basis. If a company runs a loaded artic every day in each direction between London and Glasgow then a pair of dedicated vehicles will be economic. However, if two vehicles run each day loaded in one direction only then it will be cheaper to employ a 3PL to carry the loads and leave them to find another customer who will provide them with a backload.

The other driving force for a dedicated operation is a high level of specialization. The case mentioned above of artificial lenses for surgical insertion is such an example where a dedicated storage facility almost essential. A similar example for dedicated vehicles is Abbott-Wade, a family-owned SME from Warrington who manufacture and install high-quality staircases (Figure 2.2). They run a dedicated fleet of vehicles to deliver to customers. Part of the vehicle is fitted out to transport the necessary tools and the driver must also be skilled in the installation of the staircases. Again, a dedicated vehicle fleet is all but essential.

Finally, some products may be incompatible with others. There are many complexities relating to which hazardous materials can and cannot safely be stored together, and I will not attempt to make generalizations. On a more basic level, bags of onions would not be welcome in a clothing warehouse, where their smell could taint the garments!

Shared user solutions tend to be more economical in the case of less regular requirements. If an SME needs to deliver one pallet per week to a customer 200 miles

Figure 2.2 One of six dedicated vehicles operated by Abbott-Wade

SOURCE Reproduced by kind permission of Abbott-Wade Ltd

away it would not be viable to do anything else unless there is a very specific requirement, such as security. Similarly, if the volume to be stored fluctuates through the year it would be sensible to use a shared user warehouse so that the costs can be roughly aligned to the volume of business. Minimum commitments can also be avoided – if the customer does not want a delivery this week then there is no cost if using a shared user system.

I have concentrated mainly on whether a 3PL service should be dedicated or shared user. However, there are circumstances in which an in-house operation can have a shared user element. If a small company runs its own warehouse and allows the company next door to store a few pallets for a fee, then this is the start of a shared user operation. Similarly, I gave the example of a company making daily deliveries from London to Glasgow and suggesting it might be best to use a 3PL and let them worry about finding a return load. If there is an overriding reason for using the company's own vehicle for the outbound leg, then it would be possible to market the vehicle on the return leg. There are pitfalls – the customer may delay offloading the vehicle so it is not available for the next day's work, additional insurance will be required, and operator's licence requirements are more onerous for hire and reward transport than they are for own account transport. It may, however, be a useful way of making a bit of extra money, and some major multination logistics companies such as Gefco actually had their origins in just this type of activity.

Again, each case needs to be considered on its merits.

Charging mechanisms: Open or closed book?

According to their 2018 annual report, Wincanton (one of Britain's largest 3PLs) charge 59 per cent of their business on an open book basis, and 41 per cent on a closed book basis. However, it is likely that the former will comprise mostly their larger customers and for an SME the latter is likely to be more common. But what do these terms actually mean?

A closed book tariff is what most people will instinctively consider. For example, it may be agreed with a 3PL that storage will cost £2 per pallet per week, handling £3 per pallet in, £3 out. To deliver one pallet to Manchester will cost £50, to Stuttgart £150. In many cases this suits both parties. There is no minimum commitment for the customer, paying only if a sale is made. There is a risk to the 3PL in that if they have only a small number of pallets to deliver then they may make a loss, but if they run the business efficiently they can make a good profit. A typical closed book tariff is shown in Table 2.1.

Table 2.1 A typical closed book tariff for palletized deliveries from the Manchester area

Zone		Postcode areas	Economy (2–3 days) (£)	Next day (£)	Next day am (£)	Timed (£)
1	Urban NW	M, L, WN, BB, BL, OL, SK, CW, CH, WA	45.00	45.00	49.50	59.50
2	Midlands S & W Yorkshire Rural NW	B, ST, DE, NG, LE, CV, WS, WV, DY, TF, S, DN, WF, HD, HX, HG, LS, BD, PR, FY, LA	51.50	61.50	66.00	76.00
3	London North East Bristol and Swindon Cardiff and Swansea	All within M25 plus MK, SG, OX, RG, AL, LU, HP, SL, GL, SN, BS, HR, WR, CF, NP, YO, TS, DL, NE, DH, SR, SA1-7	62.50	75.00	87.50	97.50
4	Central Scotland	G, EH, ML, KY, FK, KA, PA 1–19	62.50	75.00	92.50	102.50
5	Southern Scotland Other England and Wales	TD, DG, other mainland England and Wales	67.50	70.50	Not available	Not available
6	Northern Ireland	BT	95.00	125.00	Not available	Not available
7	Northern Scotland Offshore Islands		Rates on request			

NOTES Maximum pallet size 1,200 x 1,000mm, maximum height 2,000mm, max weight 1,000kg. Weekends and Bank Holidays Excluded. All prices exclude VAT.

A halfway house may still be a closed book agreement, but include a minimum commitment. The 3PL may agree to provide storage space for 500 pallets for three months at £750 per week whether it is used or not, or agree to provide an 18 tonne vehicle with tail-lift (including driver) for five days per week for three months at a fixed price of £400 per day including 1,250 miles per week, additional mileage to be charged at 50p per mile. This reduces the risk and the profit margin for the 3PL whilst decreasing price but increasing risk for the customer. More will be said about vehicle operating costs later.

This can be taken yet further by the principle of open book costing. In this scenario all costs are revealed by the 3PL. The customer pays these costs plus an agreed management fee, which in the case of a very large contract may be as low as 1 per cent. There is only a small margin for the 3PL but they bear little risk, as increases due to external factors (for example if the government were to increase employers' National Insurance payments) can often be passed on at cost under the terms of the agreement. A fictional example is shown in Table 2.2. In this case costs are consolidated so that space includes rental, business rates, light, heat and even security, and personnel costs include national insurance and pension contributions. In some cases these are identified separately.

Table 2.2 A typical open book tariff for warehousing

Fixed costs	Number	Unit cost (£)	Weekly costs (£)
Dedicated space (sq ft)	50,000	0.23	11,500
Manager	1	1,068	1,068
Shift supervisors	2	625	1,250
Admin staff	2	380	760
FLT drivers	12	480	5,760
Packers	6	360	2,160
Reach trucks	4	90	360
Counterbalance FLTs	2	78	156
Pallet trucks	1	55	55
System licence fee	1	1,400	1,400
Management fee	1	1,250	1,250

(continued)

Table 2.2 *(Continued)*

Variable costs		(£)
Additional storage	Per pallet per week	1.75
Admin overtime	Per hour	16.30
FLT overtime	Per hour	20.60
Agency labour		
– Admin	Per 8 hour day	144.32
– FLT drivers	Per 8 hour day	164.32
– Packers	Per 8 hour day	144.32
– Overtime	Above rates pro rata plus 33%	
Additional MHE hire	At cost	At cost
FLT gas	At cost	At cost
Consumables	At cost	At cost

In practice, things may not be quite as they seem. The 3PL may have a 25-year lease on a warehouse. If the customer offers a back-to-back 25-year commitment then the rental will almost certainly be charged at cost, but without this the 3PL may feel justified in charging a premium. Similarly, it may be agreed that vehicle repairs will be charged at x pence per mile, and the 3PL will take the risk and potential benefit of minimizing repairs, for example by reducing accident damage. Finally, it has been known for a particular cost from one of the 3PL's suppliers to be quoted at, say £10 per hour. This is the amount charged to the contract and the contract manager will be able to produce an invoice for that sum. However, at the end of the month the suppliers will send a credit note for 50p per hour to the 3PL's head office by way of a volume-related rebate. In setting up such a contract it may be worth clarifying whether the customer will benefit from such rebates.

On the other side of the fence, a customer may well seek commitments to productivity improvements over the course of a contract. It may be, for example, that the average picking rate in a warehouse is 120 items per hour, and the contract states that this should be increased to 126 per hour after one year and 132 per hour after two years. Complications may, however, arise if the customer changes the nature of the operation, for example by eliminating stipulated minimum order quantities from its customers. Such discussions can get very complicated, and genuine goodwill from both sides is necessary to make an open book contract work in the interest of both parties.

In conclusion, closed book contracts will usually be the most appropriate for smaller businesses. As the business expands, a move to a minimum commitment and then to a fully open book charging mechanism will be worth serious consideration.

The 4PL option

I have said that, in simple terms, most business can fulfil their logistics needs either by doing it themselves or by using one or a small number of 3PLs. However, there is another option known as 4PL. To once again put things in simple terms, a 4PL operation is one that controls a number of 3PLs.

This may be best explained by using an example. Let us consider a hypothetical firm in Coventry that manufactures materials for the construction industry. Due to the nature of the unloading facilities at their customers' premises materials must be delivered on a flatbed trailer (I talk about types of trailer in a later chapter but a flatbed is shown in Figure 2.3). Whilst flatbed trailers were almost standard equipment 50 years ago, these days they are generally only used where there is no viable alternative.

Figure 2.3 A flatbed trailer

SOURCE Reproduced with kind permission of Stears Haulage Ltd

It would be possible to base a large fleet of artics and flatbed trailers on site and use these for nationwide deliveries. This would, however, prove expensive as backload opportunities would be limited. A cheaper solution would be to identify 3PL companies that are making flatbed deliveries from other areas into the Midlands, and to backload these to their respective regions. A core fleet of just four vehicles might perform local deliveries on a 3PL basis within, say, 50 or 100 miles of Coventry.

To manage this operation on a day-to-day basis would take time. The options are therefore to take on more staff, or to use the services of a so-called 4PL. They would already have the necessary expertise and in many cases already have relevant contacts. This can result in a highly cost-effective solution for the company concerned, and is also attractive to the 4PL provider as they can make good profits by employing just a small number of people and without investing capital in trucks and other major assets. The 4PL idea is an anathema to some as they have a strong business instinct to 'cut out the middle man'. However, this is a case in which the middle man (or woman) can genuinely add value. It represents the best solution in only a small percentage of cases, but should not be rejected purely on the basis of business dogma.

One sometimes hears mention of a 5PL. Such an organization might combine the freight carried by several 3PLs and/or own account operators to generate economies of scale, and would have a beneficial role to play. One can also find discussion of 6PL and higher numbers on the internet, even including one definition of 10PL as 'supply chain becomes self-aware and runs itself' (Narasimhan, 2015). If your supply chain is self-aware and running itself, you will not need this book!

Conclusion

In this chapter I have looked at the various 'big picture' options for operating a logistics system: should it be an own account operation or subcontracted to a 3PL (or in some cases a 4PL), should it be a dedicated or shared user operation, and if subcontracted should it be charged on an open or closed book basis?

I have looked mainly at domestic services. The same principles apply to international movements, and whilst it would be rare for a company to, for example, charter its own ships there are examples of dedicated own account truck fleets operating internationally.

In some cases, a decision will be easy, and in others a logistics manager may receive a policy directive from board level as to which strategic direction to take. My overall recommendation is that each element should be looked at on its own merits and cost analysis conducted where necessary. In many cases a mixed approach will be best and, as a generalization, a 'horses for courses' approach is better than 'one size fits all'.

Storage 03

The physical infrastructure

In this chapter I will discuss physical storage facilities – how many are needed, and what form of racking or shelving (if any) needs to be installed. I will discuss vehicle loading and unloading and equipment such as forklift trucks in Chapter 4, operational methods in Chapter 5 and warehouse management systems in Chapter 6. Please also note that I am only discussing storage warehouses here – transhipment facilities will be discussed later.

How many storage warehouses do I need?

If you are based in the UK, the short answer is 'not many, and probably only one'.

It is true that the major grocery chains have many UK sites, but this is largely driven by scale. There are warehouses of 1 million square feet (93,000 square metres), such as those used by Sainsbury's at Daventry and Amazon in Fife, but no-one has attempted to operate a warehouse of 5 or 10 million square feet (465,000 or 930,000 square metres). Whether planning permission would be granted and whether enough staff could be recruited locally would be other obstacles.

The major motor manufacturers, by contrast, operate single UK parts centres. The Peugeot Citroen parts centre at Tile Hill in Coventry, for example, comprises 618,000 square feet (57,000 square metres) (MK2, 2014) and serves the whole of the UK. Ford has a similar site at Daventry, as does Mercedes at Milton Keynes, serving both car and truck dealers. Each has a well-planned transport network to achieve fast delivery for parts for vehicles off the road. The biggest storage warehouse I know of is the John Deere Parts Distribution Centre in Illinois, of 2.6 million square feet (240,000 square metres) (MWPVL, 2018). John Deere is most famous as a tractor manufacturer, but also markets other equipment for the agricultural and construction industries.

The driver behind this is the nature of their inventory. There are tens of thousands of line items held by each manufacturer – John Deere at the extreme hold over 800,000 different parts in Illinois. This occurs because there are a great many different parts on a vehicle, and a great many vehicle variants, including those no longer in manufacture. A few will be relatively fast moving, such as brake pads for current models, but most will be called off but rarely and will require just a small stockholding. To duplicate a small stockholding of tens of thousands of parts would attract a prohibitive cost, both in terms of space and in terms of the capital tied up in the inventory.

For other businesses, there are likely to be other drivers towards a small number of locations. For a small or medium sized company, it will be intuitively the right solution to have just one site. In some cases the 'warehouse' may be called the 'stores' and be contained within a room no bigger than a large office. To duplicate a small operation would all but double the cost. A single location for all the stock is in practice much easier to manage than several, and there is also the psychological and practical advantage of having all one's stock close at hand. Whilst fast delivery to customers can be important, this is often best achieved by an efficient transport network rather than by regional stockholdings. It is quite feasible to guarantee 8am next day deliveries except for geographical extremes such as the north of Scotland. An important issue is that in multiple warehouses stock of a certain item can run out in one but not in others, thus causing a customer service failure which would not have occurred using a centralized warehouse.

There are few markets that demand that items ordered are delivered within the same working day. Those that do would be exceptions to my general conclusions, and would require local facilities to achieve this.

Having said all of this, some would argue that it would not be good business practice to make such a decision without a proper financial analysis, and it is difficult to disagree with such a view.

Figure 3.1a and 3.1b illustrate this analysis for hypothetical cases.

Let us consider the example of a fictional wholesaler. Goods are brought in to Felixstowe by container and sourced from UK suppliers; brought to one or more storage warehouses; stored until needed; then distributed throughout the UK, roughly in line with population density. We assume for the purposes of this exercise that this is an established, stable business, and that we are working from a clean sheet – let us say, for example, that the lease on our current warehouse is coming to an end. Our objective is to work out the optimum number of warehouses to operate. Options might include a single warehouse in a prime location such as Milton Keynes or Lutterworth; one warehouse in Milton Keynes and a second on the M62 corridor; four or five to include regional distribution centres in Scotland and the South West; or 10 facilities including one in Norfolk and one in Devon.

The major difference between these two examples is the nature of the business. In example 1 (Figure 3.1a), full load vehicles are received and deliveries are in full pallet quantities; in example 2 (Figure 3.1b) pallets are received, sometimes in full loads, and broken down so that shipment to customers is in parcel quantities.

Figure 3.1a Cost variation by number of warehouses: full loads in, pallets out

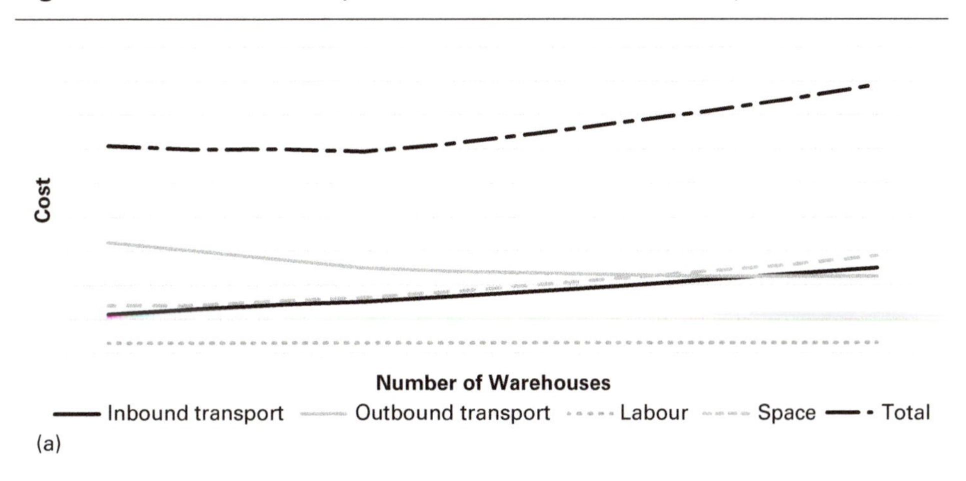

Figure 3.1b Cost variation by number of warehouses: pallets in, parcels out

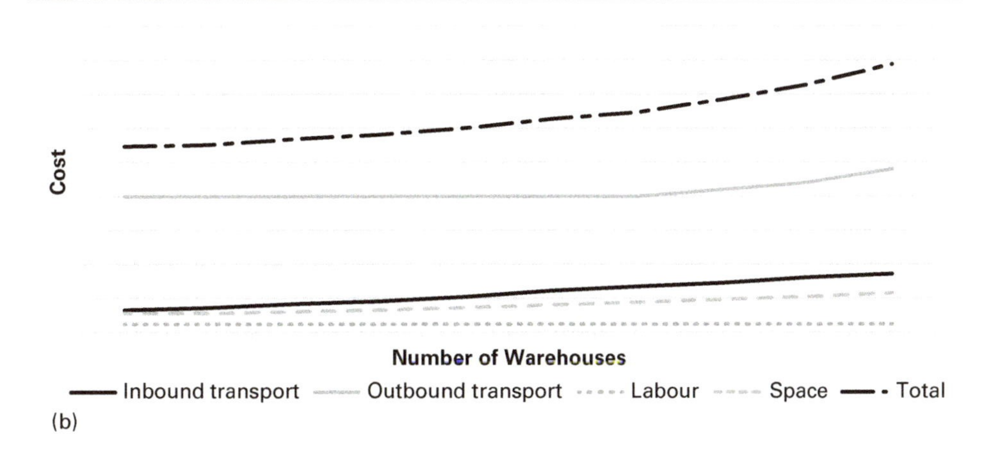

There are five main cost factors, of which four are illustrated on the graphs.

1 Inbound transport. In example 1, if there are a small number of sites all material will be delivered in full loads. As the number of sites increases some will be more distant so prices will increase – road haulage of a container to Milton Keynes may cost £400, to Leeds £650 and to Glasgow £1,200. As the number increases yet further, it will be necessary to deliver in less than full load quantities which will increase costs significantly. This is even more marked in example 2, as deliveries of small numbers of pallets to numerous sites including Devon and Norfolk will be expensive.

2 Outbound transport. For pallet deliveries, this will decrease as the number of sites increases. Deliveries to Scotland from a site adjacent to the M8 between Glasgow and Edinburgh will be cheaper than such deliveries made from anywhere in

England. However, this is unlikely to be the case with parcels deliveries. Collection of 1,000 parcels from a Midlands distribution centre will be more attractive to a parcels carrier than collection of 100 parcels from each of 10 warehouses including one in Devon, and costs of outbound parcels transport are more likely to increase than to decrease as the operation is further subdivided.

3 Labour. In theory, labour cost should increase in a larger facility, as operatives have further to travel to put away and pick items. However, in a larger facility there are in practice economies of scale and the two factors will often tend to cancel each other out.

4 Space, in which I have included overheads. This is where economies of scale really become important. Larger facilities tend to have greater free heights and whilst the cost per square metre may be constant the cost per cubic metre will be reduced, and cost per pallet stored will fall correspondingly. The ability to hold full pallet quantities at a single site, but less than full pallet quantities at a number of sites, will also be important, as will the possibility of using space-saving storage methods such as drive-in racking (see below) which are better suited to larger quantities of goods. Overheads are also reduced in relative terms. To take an extreme example, the general manager of a 500,000 square foot (46,000 square metre) warehouse will attract a higher salary than each of the 10 managers of 50,000 square foot warehouses, but it will not be 10 times the salary.

5 Inventory. I have not attempted to illustrate this, as the result is too variable. If an organization holds, say, 1,000 pallets of each of ten line items, there will be no difference in the inventory cost whether these are spread across one, two or ten sites. However, in cases where it is necessary to hold a minimum quantity at each location in order to service customers, the inventory cost will rise dramatically. It is important to remember that whilst purchasing extra inventory does not directly affect profit and loss, the interest payable on capital borrowed to buy that inventory does, and cash flow is especially important to smaller businesses.

To perform the analysis, it will be necessary to look at the data for your business for a typical period. Indicative quotations will need to be obtained from suppliers and the above costs compared on spreadsheets. This will be a time-consuming process, but to be fair it will be good business practice.

Taken overall, for all but the largest corporations a single storage facility is likely to be the most cost-effective solution for serving the UK, or at most a small number. Many multinational corporations come to a similar conclusion regarding continental Europe, with a single European distribution centre often located in Belgium or the Netherlands. It would be unusual for many regional depots to be the most cost-effective solution.

Storage options: Block storage, racking and shelving

Once you have decided on the number of warehouses, you will need to work how to operate them.

As a first step in this process you will need to work out how to lay out the building – will you use racking, shelving or neither? In the following sections I have made an implied assumption that this will be setting up an in-house operation, but the information will also be useful to those sourcing a third party solution as it will give a better idea of what to look for – although it should be borne in mind that a 3PL will most likely lay out their warehouse to be flexible and meet the needs of a variety of potential customers. I have also restricted discussion to conventional and semi-automated systems – fully automated systems will be dealt with in a later chapter.

The short answer is that pallets usually need racking, and cartons usually need shelves, but things are not quite as simple as that.

Outdoor storage

The cheapest form of storage is outdoor. The only items for which this is really an option are those designed for outdoor use. Cars are an obvious example. About 200 per acre can be stored in a large compound, and to try to store them indoors would be prohibitively expensive. All they need after outdoor storage is a good clean and they can be delivered to dealers. Outdoor storage can also be used for many items used in the construction industry and some bulk items such as coal or scrap metal.

It might also seem obvious that items not designed for outdoor use should not be stored outside, but this is sometimes forgotten. Many companies are tempted to keep returned goods outside; however, they will deteriorate quickly and any hope of salvaging residual value is quickly lost.

Similarly, metal stillages and wooden pallets can often be seen stacked in yards. They inevitably get wet and dirty, and will damage goods placed on them – a box with a wet or water-stained base will render an item unsaleable. Security is also an issue. A man was convicted some years ago of selling pallets to local companies, and returning at the weekend to steal them from respective yards to resell once again!

Block storage

Not all pallets need racking. Block storage is the industry term for keeping pallets in stacks on the floor (Figure 3.2). The obvious advantage is that no investment is required and pallets can be stacked more closely than in any form of racking. Up to 80 per cent space utilization can be achieved.

Figure 3.2 Palletized goods in block storage

SOURCE Reproduced by kind permission of SEP Logistik AG

However, in practice block storage is suited to few products. The product and/or its packaging need to be stackable: to have just one or two layers of pallets on the floor of a 6.5 metre high warehouse wastes an awful lot of space. Suitable products include car parts in metal stillages, which have been used in the automotive industry for decades, or any product in more modern plastic reusable packaging. Canned drinks and forest products may have the strength in themselves to allow stacking. It is important to ensure that there is no danger of collapse – a tipping stack is a major health and safety issue.

Perhaps less obvious is that this method is only effective if there are a large number of pallets of each stock item. New banknotes would be an example – the UK has just four denominations in circulation and there is no need for any first-in, first-out rotation, so the metal stillages in which they are stored can be tightly packed into a vault. If, however, a business has a few pallets each of a large number of part numbers then the two options are to have a large number of small pallet stacks, which is wasteful of space, or to stack different stock items together, which means having to move most of a stack to get to the bottom pallet then reassembling the remainder. In practical terms this will be a nightmare.

Racking: Some words of warning

We have all heard tales of people who have wanted to buy an item, and simply selecting the cheapest they can find on the internet. The results can be relatively minor, such as a leaking printer cartridge, or fatal in the case of people buying fake drugs.

There are sadly people who sell racking on the internet that cannot be guaranteed as safe. Some suppliers are even unable to say where it is manufactured. Often it will look as though it is fit for purpose, and initially work well. However, it may in practice be vulnerable to small amounts of damage, and fail as there is no built-in safety margin. True responsibility is often difficult to pin down, as the supplier will blame the damage for any collapse: even the best racking cannot sustain major damage. This does not of course apply to everyone who sells racking online, as the most reputable suppliers do so also.

I would therefore strongly recommend that anyone installing racking ensures that it complies with BS EN 15512, and that the company supplying it is a member of the Storage Equipment Manufacturers Association (SEMA). SEMA issue a code of practice incorporating high standards for health and safety, with which all members must comply. These standards were introduced in 2008, with minor changes in 2014, and there is a consultation process under way with the aim of issuing a new more rigorous design code in 2019. Members are subject to an independent annual assessment carried out by the University of Salford, and individual installers should carry ID cards to show they are members of the Storage Equipment Installers Registration Scheme (SEIRS). Using such suppliers will ensure that the racking is of a high standard for both durability and safety.

Also, any racking has a maximum safe working load. Typically, this will be 1 tonne per pallet space, which will be sufficient for most products. If your product is heavy, such as tinned fruit or vegetables, a higher safe load should be specified. It is possible to supply racking with a capacity of up to 30 tonnes per bay (a bay is the area between two uprights, including all levels), and to narrow the bays to a single pallet width. This will be sufficient for all but the heaviest products.

If storing a mix of products, procedures must be in place to ensure that the heaviest products are not stored together – a procedure that specifies they are kept at floor level may be the easiest way to achieve this. An example is mixed automotive parts. Most of these are light, but there are a few exceptions, such as car batteries. If pallets of these are all kept at floor level, there will be no problem. However, if someone were to decide to keep all the pallets of car batteries in the same bay the result might be a very dangerous collapse of the racking.

The final word of warning relates to floor loading – ie the strength of a building's floor. If you are considering renting or buying a warehouse, ask what the maximum floor loading is. The response may be of the nature, 'This is a modern building built to the latest standards', which is meaningless, or it may be of the nature, '35 kilonewtons per square metre', which I must admit is equally meaningless to most. It means that for each square metre of space the floor is designed to take an evenly distributed load of just over 3.5 tonnes. Some modern warehouses of up to 10 metres high are being built to this specification, and it would be insufficient for an evenly distributed stack of a moderately heavy product such as beer, even to half the height of the building. If you wish to install racking to utilize the full height you will have serious problems: the load on each post can reach 80 kilonewtons and this is not evenly distributed over the floor. It is not possible to generalize and cover all eventualities, and I would strongly advise anyone considering building or renting a warehouse with the intention of installing racking to consult with a reputable manufacturer or distributor to confirm suitability before entering into a legally binding commitment.

Standard wide aisle racking

This is the type of racking most commonly in use in smaller warehouses, and the type with which even most laymen are familiar (Figure 3.3). Sometimes also known as adjustable pallet racking or APR, it is more flexible than any other type of racking.

The most obvious advantage is that each pallet has its own location and can be accessed directly. Whether there are one or two pallets of each item or a thousand of each, the racking can be operated equally well. In practice, it is the easiest type of racking to operate and speed of throughput will be faster than with any other type. Usually the aisles are about 3 metres wide, which means that standard reach trucks can be used, avoiding the need for expensive specialized narrow aisle equipment

Figure 3.3 Wide aisle racking

SOURCE Reproduced by kind permission of Redirack, a trading division of Stakapal Ltd

(types of forklift are discussed in Chapter 4). In an even wider aisle layout, with 3.5–4 metre aisles, counterbalance trucks can be used – this may be a benefit in a small warehouse as it avoids the need for more than one type of forklift.

Requirements such as first-in, first-out can be easily controlled using any good modern warehouse management information technology (IT) system and a proactive warehouse manager will ensure that fast moving items are stored in the most easily accessed locations.

Also, the racking can be adjusted. If it has been installed with pallets of a certain height in mind, but requirements change, it is a relatively easy matter to move the cross-members up or down accordingly.

The main disadvantage is that the wide aisles result in poorer space utilization than other racking types – around 40 per cent is average (Redirack, 2018). It should be noted that all such figures I quote can only be approximate. The dimensions of a warehouse may just allow an extra row or just prevent it, and roof support stanchions can be found to be in the most inconvenient of places when one comes to plan a detailed racking layout.

Narrow and very narrow aisle racking

To make better use of space, narrow aisle racking can be installed: this has been the biggest growth area in racking installation in the last 20 years. This is similar to conventional wide aisle racking but, as the name implies, the aisles are narrower! In theory, aisle widths can be as low as 1.6 or 1.7 metres, but at these minimum widths operation becomes more difficult (and slower) and it may be better in practice to use a slightly less narrow aisle of up to 2.3 metres. This type of racking gives better space utilization, typically 45 per cent (Redirack, 2018). The main advantages of wide aisle racking still apply – each pallet is accessible, it is suitable for large numbers of different items, and first-in, first-out operation can be easily managed. It is still fairly easy to adjust the racking if required.

The main disadvantage is that specialist narrow aisle forklifts must be used. Options include turret trucks, and the articulated type. These are more expensive than standard reach trucks, so a higher initial investment is required. They are also more difficult to operate so good operators need to be recruited and retained.

Very narrow aisle racking is packed even tighter, with aisles as narrow as 1.4 metres, although exact definitions vary. Highly specialized forklifts must be used, which may rely on guidance by wires sunk into the floor. These often have man-riser cabs, in which the driver is raised to a similar height to the forks, and require a high level of expertise to operate effectively (Figure 3.4).

Figure 3.4 Very narrow aisle racking. Note specialized fork-lift with driver raised to level of forks

SOURCE Reproduced by kind permission of Redirack, a trading division of Stakapal Ltd

Drive-in, push-back, pallet flow and double-deep racking

In the earlier section on block storage, I said that this was best suited to pallets that are stackable, where case or piece picking from pallets is not required, and there are few different stock items. In some cases, the latter two will apply, but the pallets are not stackable.

In these circumstances drive-in racking may be the best answer. In this configuration pallets are not all stored accessible from the aisles, but are stacked in front of each other, perhaps five or 10 deep. The racking is configured so that forklifts can drive right into lanes in the racking to access the pallets at the rear: often there are lead-ins at the entrance to each lane and guide rails along the sides to assist: specialized forklifts are needed (Figure 3.5). Space utilization is much greater, with up to 85 per cent floor coverage.

However, operation is much more difficult and less flexible. The pallets at the rear can only be accessed if the spaces in front are empty on all levels – it is not possible to get to the back of one level if a different level is full. As a result, percentage utilization of the theoretical number of spaces tends to be lower than with wide or narrow aisle racking. First-in, first-out operation is difficult, case picking from pallets is impractical, and there is little flexibility.

Figure 3.5 Drive-in racking

SOURCE Reproduced by kind permission of Redirack a trading division of Stakapal Ltd

Push-back racking is similar, except that a forklift does not drive into the racking, but pushes pallets back into the racking by placing another in front of it. There will be a moveable frame within the racking to allow this, and when a pallet is removed the pallet(s) behind will roll slowly forward. Pallets can be stacked up to four deep using this method. Its main advantage over drive-in racking is that specialist forklifts are not required.

A variation is pallet flow racking (Figure 3.6). Again, there are often five or six pallet spaces in front of each other, but on each level is a sloping set of rollers. Pallets are fed in at the rear, and slide forward under gravity to the front. Pallets are then drawn from the front as required. This is again best suited to relatively few stock items, and is not easy to adjust. A high-quality pallet is needed – a collapsed broken pallet will cause a lot of disruption. It is also more expensive to install. However, it has the advantage over drive-in racking that the front pallet can always be accessed – it does not matter if one level is more full than another. First-in, first-out is also much easier to manage.

Another option is double-deep racking, in which the racking is in a conventional layout but there are two pallet spaces in each location, one in front of the other. This gives some of the best but some of the worst of both worlds – there may be 50 per cent space utilization (Redirack, 2018), but selectivity is reduced by 50 per cent. Specialist forklifts are again required.

Figure 3.6 Pallet flow racking showing pallet rollers

SOURCE Reproduced by kind permission of Gonvarri Material Handling

All these types of racking installation are in long-term decline as shuttle racking increases in popularity (see below). Double-deep racking in particular is often seen as a compromise that offers neither full flexibility nor optimum space utilization, and is but rarely used.

Semi-automated racking

The are two types of semi-automated racking in use – shuttle racking and mobile racking.

The former is similar in principle to push-back racking, except that pallets are moved by a pallet shuttle (Figure 3.7). This is a motorized platform that fits on runners in the racking. A pallet is placed on it by a forklift operator, who will be carrying a remote-control handset. He or she will use this to move the platform back in the racking, then lower it slightly to allow the shuttle to return to the front of the racking, and repeat the process. The reverse process is used for pallet retrieval.

The shuttle is electrically powered and usually has a rechargeable lithium battery – typically they can work for 12 hours between charges. Most will have an automatic low battery function to return the shuttle to the front of the racking. It can be moved by forklift between runs, and it is prudent to have more than enough shuttles available in case of increased demand or mechanical breakdown.

Figure 3.7 Shuttle racking showing empty pallet shuttles

SOURCE Reproduced by kind permission of Avanta UK Ltd (www.avantauk.com)

It is possible to install pallet runs of 50 metres, which permits excellent floor utilization. First-in, first-out operation can be achieved by accessing the pallet runs from both ends. This system has several advantages over drive-in racking. No specialized forklifts are needed, which partly offsets the higher cost of shuttle racking, and there is less risk of damage as forklifts are not driven into the lanes. It is easier to operate and throughput is therefore quicker, and shuttle racking is therefore tending to replace drive-in racking in the marketplace.

In a mobile racking installation, it is the whole run of racking that moves (Figure 3.8a–d). A specialized installation is required with rails set into the floor, which may itself need strengthening. The runs of racking are controlled remotely, and can be moved together or apart so that an aisle is created only between the particular runs which require access at that particular time. All the advantages of conventional wide aisle racking apply, with access to every pallet available, but double the space utilization can be achieved.

The main disadvantage is that mobile racking is expensive – in most situations it will be more cost-effective to install a non-motorized system and occupy more space. The exception is where the space itself is expensive, and mobile racking is therefore most commonly used in chilled or frozen food storage.

Figure 3.8a Sequence of four pictures of a mobile racking installation showing aisle in centre closing and aisle on right opening as the racking span is moved. Note electrical control buttons

Figure 3.8b

Figure 3.8c

Figure 3.8d

SOURCE Reproduced by kind permission of Storax

Specialized racking and accessories

The options above are suitable for most types of palletized goods. However, some goods need a specialized solution.

In many cases, standard racking can be used by fitting accessories. If items smaller than a full pallet need to be stored, then there are several options. These include a

Figure 3.9 Cantilever racking used for storage of metal tubing

SOURCE Reproduced by kind permission of Avanta UK Ltd (www.avantauk.com)

half-pallet insert, which as the name implies allows safe storage of a half-sized pallet. Wooden decking, wire mesh shelves and steel sheet shelf panels are also available and are suitable for undersized pallets, cartons or other smaller packaging units.

Perhaps the most common form of specialized solution is cantilever racking (Figure 3.9). This has no continuous uprights at the front of the racking, but simply horizontal bars reaching forward from the rear uprights, which are usually adjustable as with pallet racking. Cantilever racking is designed to accommodate long items that will not fit into conventional racking – examples include pipes, timber, rolls of carpet and sheet metal.

An alternative is vertical racking, designed to allow the vertical storage of long items – whilst this would not be recommended for pipes due to the risk of damage to the bottom end, it is commonly used for timber.

Finally, some products require racking specially designed only for that product. Hanging garments, tyres (Figure 3.10) and coils of rope or cable are examples.

Other considerations for racking installations

When costing or specifying a racking installation, it is not only the racking itself that needs to be considered.

Figure 3.10 Specialized racking used for storage of tyres

SOURCE Reproduced by kind permission of Gonvarri Material Handling

First, the racking will need protection. In any forklift operation there will inevitably be some collisions with the racking and it is therefore prudent to protect at least the ends of the pallet runs with barriers, and the uprights with column guards. Also, whilst runs of racking are often back-to-back, this is not always the case, especially at the edge of the racking area. If there is a risk of items falling from the back of a pallet space, then protective mesh should be installed to prevent this – the health and safety implications of not doing so are obvious. Another health and safety requirement is to display load notices that give the maximum safe load of each pallet space and bay. All these items can be provided by a racking supplier.

An additional major expense is lighting. Many warehouses have lighting that is quite adequate when only the floor is used. However, if racking is installed to almost the full height without additional lighting the aisles will be too dark to operate safely or effectively. Additional lighting must be provided along the aisles, and it will in practice be easier to do so before the racking is installed – this should be remembered at the project planning stage.

Access between aisles for forklifts is often provided via the handling area at the end of the aisles. However, if the racking layout precludes this then a wide lateral aisle is necessary. This can take the form of a tunnel with racking at the higher levels to allow at least partial use of the floor space. A pedestrian tunnel at the closed end of the aisles may also be necessary to permit access to fire exits, which of course should be clearly signposted from within each aisle.

Finally, racking will require labelling – options for this are discussed in Chapter 5, but it should again be remembered at the project planning stage that labelling a large number of pallet spaces is not a quick job.

Smaller items: Shelving

I have discussed at length the various types of racking systems available for pallet storage. However, smaller items are best stored on shelves. This may include cartons that are shipped complete; outer cartons from which inner cartons are picked; cartons from which individual pieces are picked; or items that are stored loose.

Standard shelving resembles miniature racking, with steel uprights and shelves between rather than cross-members to support pallets (Figure 3.11). It is available in a wide variety of shapes and sizes: heights of up to 12 metres; shelf depths (front to back) from 30 to 120 centimetres; shelf widths (side to side between uprights) of up to 3m; and load capacities of up to 900 kilogrammes per shelf. The shelves themselves can be of chipboard, wire mesh, stainless or painted steel, or melamine covered steel. For most products, selecting the right type of shelf will simply require determining weight and dimensions of the business's products and selecting the appropriate shelf length height and width and load capacity.

Loose items may be prone to rolling around on shelves, which will cause problems when they end up in the wrong place. It is possible to subdivide shelves using full-height dividers. These are steel sheets that are fitted vertically between the shelves – this solution might, for example, be used for spare parts. An alternative is

Figure 3.11 Standard shelving used for storage of a variety of goods

SOURCE Reproduced by kind permission of Avanta UK Ltd (www.avantauk.com)

Figure 3.12 Plastic storage bins used for storage of small items

SOURCE Reproduced by kind permission of Gonvarri Material Handling

Figure 3.13 Carton flow shelving used for storage and picking of confectionery

SOURCE Reproduced by kind permission of Avanta UK Ltd (www.avantauk.com)

to use a plastic box, either with or without lid, of the type that is used in many homes for storing children's toys, for example. For very small items such as nuts and bolts purpose-designed plastic storage bins, with a lowered front edge, are usually the best solution. These are illustrated in Figure 3.12.

Carton flow shelving, also known as carton live storage, is similar in principle to pallet flow racking (Figure 3.13). Cartons are placed on shelves, which are usually fitted with roller wheels and angled forward (though some types have simply low-friction angled shelves), so that as the first carton is removed from the front other cartons roll forwards. This allows for an efficient picking system, often with separate personnel replenishing from the rear. This system is well suited to situations in which a high volume of piece or inner carton picking is required, and has a valuable role to play in such applications.

Finally, mobile shelving, of a similar type to mobile racking, is also available. It is usually manually moved rather than being motorized, and accessing items will be time-consuming. It is best suited to circumstances in which a large number of items must be kept, but each is accessed very rarely, such as spare parts for models of car no longer in production.

Mezzanine floors

The original definition of a mezzanine floor was as an intermediate floor between the main floors of a building. In the warehousing context it has come to mean a semi-permanent steel floor erected to make better use of vertical space (Figure 3.14).

There are two reasons for possibly installing a mezzanine. First, there may be space that could not otherwise be used, such as above a single-storey office. A mezzanine

Figure 3.14 Mezzanine floor with shelving on both levels

SOURCE Reproduced by kind permission of Avanta UK Ltd (www.avantauk.com)

will allow shelving to be installed in that space and/or for it to be used for activities such as rework of returned goods.

Second, and more commonly, the need arises because it would be wasteful of vertical space to install shelving of, say, 2 metre height but not use the space above. Whilst there are methods of picking from high racking (these will be discussed later) it is more efficient, and faster pick rates will be obtained, if operatives remain at ground level.

This is best achieved by installing one or more mezzanine floors. Shelving can be installed at ground level and on each level above – if space is at a premium, mezzanines can be installed with low ceilings to minimize space lost above shelving runs. Whilst it is most common to install just a single additional floor, several can be installed if necessary.

Mezzanines can be combined with pallet racking. One example is that used by a well-know fashion retailer. Goods were received on pallets but picking for delivery to stores was on a single item basis – maybe a single bracelet or item of underwear. Racking was arranged in single runs, and was replenished from the rear by conventional reach trucks. On the other side there was a mezzanine floor installed at every second racking level and operatives moved along the levels picking items into plastic tote boxes.

Whilst operatives can walk up and down to a mezzanine, it is not feasible for them to carry goods up and down stairs except on a very limited and occasional basis. A system of conveyors and/or a goods lift will therefore need to be installed.

How big does my warehouse need to be?

In the ideal world, you would be able to consider the best option for racking or shelving, and know the exact number of pallets you will be storing. This can then be used to calculate the ideal size of warehouse, and you will be able to find a warehouse of exactly the right size available at short notice in exactly the right location, at a very cheap price.

Unfortunately, this rarely if ever happens in the real world, and sometimes decisions have to be made on a 'chicken and egg' basis. You may be locked in to a long lease on an existing facility, which may already be fully racked, and in extreme cases the amount of stock to be held may even be driven by the available space rather than vice versa. Even if setting up a new warehouse, few companies can afford to design and build a warehouse to their own ideal specification and commit to a freehold purchase or 25-year lease. In practice, decisions will be driven by what is available on the market in the right area at the right time, and cost will always be a factor.

Having said this, you will need to know what to look for, so here are some guidelines.

Racking

As a rule of thumb, each 1,000 square metres of warehouse space will be able to accommodate standard wide aisle racking sufficient for 270 standard 1,200 x 1,000 millimetre pallets at ground level. (This cannot be an exact formula as buildings are almost always a bit too narrow or a bit too wide for the best width of aisle, and stanchions and fire exits can be found to be in the most inconvenient of places.) We can use this to calculate the capacity for each 1,000 square metres of space, according to the number of levels of racking. This is summarized in Table 3.1.

These figures can be pro-rated for larger warehouses. It should be remembered that to try to operate a warehouse at 100 per cent capacity will be inefficient, but to operate at, say, 50 per cent would be wasteful of space. Utilization of 90–95 per cent is the best compromise and this should be taken into account in space calculations.

Shelving, specialized storage and rework areas

Unfortunately, there are a great variety of sizes and configurations of shelving, and no rule of thumb is available. The only way to calculate shelving space requirements is to calculate from scratch based on the number and size of spaces available. The same principle applies to secure areas, cantilever racking, and other types of specialist storage.

Table 3.1 Pallet capacity per 1,000 square metres of warehouse space

	Type of racking				
Levels of racking	Wide aisle	Narrow aisle	Double deep	Motorized	Drive in (full depth)
1	270	300	335	500	570
2	540	600	670	1,000	1,140
3	810	900	1,005	1,500	1,710
4	1,080	1,200	1,340	2,000	2,280
5	1,350	1,500	1,675	2,500	2,850
6	1,620	1,800	2,010	3,000	3,420

SOURCE Richards (2018), Redirack (2018) and author's own calculations
NOTE Based on 1,200 x 1,000mm pallets, assuming 100% utilization. These figures are a guideline only and a site survey should be undertaken before any decisions are made.

Handling area

For a typical storage warehouse using dock levellers (see Chapter 4) and/or loading in the yard, and with stock turnaround every 2–3 months, 10 per cent of the space should be set aside for shipping and receiving. This should be considered as a minimum, and for fast-moving warehouses or those with very dense storage this figure will need to be increased – up to 100 per cent of the working space for a cross-dock facility.

Offices and crew rooms

One of my pet hates is to see a portable cabin taking up valuable working space in a warehouse, and providing a far-from salubrious working environment for office staff, whilst purpose-built offices remain unused elsewhere on the site. In many cases some reshuffling may be needed (perhaps the accounts department moves upstairs, for example); and there may need to be some building works to provide extra doorways between warehouse and office, or maybe a window or counter at which drivers can report. However, wherever possible offices, locker rooms, canteens, etc should be kept away from working areas, and if selecting a new location it is wise to ensure that there are adequate purpose-built offices provided.

Summary

Let us take a simple hypothetical example. You require racking space for 5,000 standard pallets; a 200 square metre area for reworking returned goods; and a secure area of 100 square metres for high value goods. Stock turn is four times per year.

An agent informs you that there is a unit available for rent. It has 6,000 square metres of warehouse space, 600 square metres of office space, and an eaves height of 8 metres. You calculate that this will enable five levels of racking given the pallet height of your particular product.

Space required will be:

Racking 5,000/1215*1000 =	4,115m^2 (from Table 3.1, assuming 90 per cent utilization)
Rework area	200m^2
Secure storage	100m^2
Handling	500m^2
Offices in working area	Nil
Total	4,915m^2

The available unit will therefore have more than sufficient space, and capacity for future growth.

Temporary warehousing

One option that is surprisingly rarely used is temporary warehousing, which can be a cheap, quick and flexible method of providing additional space.

It can be provided very quickly in an emergency – a 1,000m^2 unit could be erected within a week if necessary. Events such as fires and floods are thankfully rare, but if you did suffer such a misfortune this could get you out of trouble. I used this solution some years ago when a warehouse for which I was responsible was nearing the end of its lease in a few years' time. We were not planning to renew the lease and intended to move to larger premises at its expiry, and a temporary structure provided much-needed extra capacity during that period. Other circumstances that favour a temporary solution include uncertainty as to future needs (perhaps a new product is being launched), a lack of available capital for investment in a permanent warehouse, or simply the lack of availability of suitable space in the area.

Sizes can vary from 100 to 10,000 square metres with up to 6.4 metres height, and construction typically consists of an aluminium frame and PVC walls. Additional insulation can be provided if required, as can a range of accessories such as roller shutter doors, fire doors, lighting and enhanced security. A firm base (asphalt, concrete or compacted gravel) is needed, and if you are intending to erect a temporary warehouse you will have to apply for planning permission. The website of Herchenbach (2018) gives indicative prices of £54,000 for a basic 1,000 square metre unit, rising to £87,000 for a fully insulated version. To this will need to be added about £20,000 for erection and disassembly, £3,500 for delivery and the cost of accessories. These costs, however, can compare favourably with the cost of renting a unit of similar size.

Conclusion

In this chapter I have discussed the physical infrastructure needed for storage and warehousing of material. I have given guidance on the number of warehouses likely to be required (probably only one) and described the various options available for storage such as types of racking and shelving.

In Chapter 4 I will discuss the types of equipment available to handle that material.

Warehouse handling equipment 04

In previous chapters I discussed the number of warehouses that might be needed, and the various options for ways to store goods, such as racking and shelving. In this chapter I will discuss the various types of equipment that are available for handling those goods. This includes static items such as loading docks, and mobile equipment such as forklifts and order pickers. Together the latter are known as materials handling equipment, or MHE.

Vehicle loading and unloading

In some cases, minimal equipment is needed: your warehouse or stores operations may handle only small items weighing a few kilogrammes each. These can easily be handled manually, and no mechanical equipment is needed. A sack barrow or a simple trolley may make life easier and these can be purchased for less than £100 or £200–300 respectively (as at 2018), but there is no need for expensive MHE.

There are some locations in which heavy items are received only occasionally. Most car parts, for example, are suitable for manual handling – body panels may be awkward but they are not heavy. However, there are a few exceptions such as engines and gearboxes. There are also some businesses that receive goods on pallets, but break these down on arrival into individual parcels; and others such as security installation providers that normally despatch only parcels but occasionally need to ship a pallet or crate for a major project. Such businesses could keep a forklift on standby, but it might only be used a few times each week, and this would be an expensive option.

One alternative would be to insist that all deliveries and collections are made on vehicles that carry loading/unloading aids, such as a tail-lift or forklift truck. An example of the latter is shown in Figure 4.1.

Figure 4.1 Articulated vehicle carrying forklift truck for unloading

SOURCE Reproduced by kind permission of Pets at Home Ltd

However, not all customers and suppliers have such vehicles available, and 3PL companies often charge a premium for their use. The good news is that simpler and cheaper alternatives are available. Pallets can be moved on the level using a hand pallet truck, also known as a pump truck (Figure 4.2). These are useful even in a warehouse with mechanized MHE, and again can be purchased for just a few hundred pounds. I would recommend that all but the smallest warehouses have at least one on site.

Figure 4.2 Hand pallet truck

SOURCE Reproduced by kind permission of Linde MH

Perhaps less well know is the availability of manual pallet stackers. These are pedestrian controlled, and the forks are raised and lowered manually either hydraulically by pumping a lever, or by turning a handle to operate a winch. Various options for lift height (up to 3 metres) and load capacity (up to 1 tonne) are available. An electric motor is an optional extra. Lighter duty versions of these stackers are available for less than £1,000, so if you only need to lift pallets or crates on an occasional basis then this is probably the most cost-effective option.

One important thing to note is that the type shown in Figure 4.3 is suitable for open-boarded pallets (such as Euro-pallets) only – ie on at least two sides there is not a continuous wooden slat along the floor. If used with a fully boarded (eg Chep) pallet the raising of the forks will cause the top of the pallet to lift whilst the legs will keep the lower slat on the ground, most likely wrecking the pallet. This problem can be overcome by using a straddle leg variant with legs further apart so that the pallet can be lifted from between the legs.

Floor loading or loading dock?

Floor loading

For loading and unloading vans, the most obvious solution is to do so from floor level. The same applies to vehicles equipped with cranes or tail-lifts.

For larger vehicles with side access this remains a viable option: the curtains can be opened and a forklift used to offload or load pallets, for example. The main advantage is flexibility – the floor height of the vehicle trailer is almost irrelevant. If loading a mixed load with many different shapes and sizes of packages, it will in

Figure 4.3 Pedestrian guided and pump-operated pallet stacker

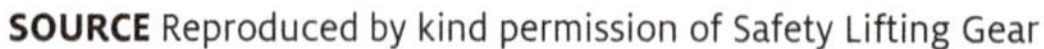

SOURCE Reproduced by kind permission of Safety Lifting Gear

practice be possible to make more complete use of the volume available on the vehicle by side loading, but these types of loads are rare – most companies ship pallets or stillages of regular sizes and shapes.

The main disadvantage is use of floor space. It is not uncommon to see vehicles being unloaded outside in the yard, but this may not be such a good idea in bad weather. If the vehicle is backed in through a ground level door (assuming there is one of sufficient height) it will require a lot of space, especially as it is best to operate from both sides.

Fork extensions

If unloading from one side only, normal length forks will not be able to reach pallets on the far side. This problem can be overcome by using a forklift with permanently fitted extended forks, or alternatively temporary fork extensions can be fitted. These are metal sleeves which can be slipped over the permanent forks and locked into place, and can be used to extend fork length, the recommended maximum additional length typically being 66 per cent of the original fork length. However, this in turn creates an additional problem that the lift capability of a forklift is much reduced if the load is not as close to the body of the forklift as possible.

> For example, the Jungheinrich TFG 320 has a lift capacity of 2,000 kilogrammes if the load centre is 500 millimetres from the mast, but this reduces to 1,300 kilogrammes if the load centre is 1,000 millimetres distant (Jungheinrich, 2018). For these reasons it is better to access vehicles from both sides, and use extended forks only for their main purpose of handling oversized pallets.

Fixed docks

Some older warehouses have fixed level concrete docks. These may be useful if vehicles of the right height conveniently need to be unloaded, but a 1.4 metre fixed dock is not a good solution for operating trailers with a 0.9 metre deck height. In the interests of future-proofing solutions as much as reasonably practicable, I would rarely recommend fitting fixed concrete docks. If you happen to have such docks, then bridge plates are available to span the gap between the dock and the vehicle as illustrated in Figure 4.4. It is also possible to fit permanent hinged metal plates for the purpose.

Figure 4.4 Bridge plate for use between fixed dock and vehicle

SOURCE Reproduced by kind permission of Metalmec SRL and The Ramp People

Container ramps

Having said all of this, some vehicles require end-loading (or unloading). If you are importing goods from the Far East they will almost certainly arrive in shipping containers that can only be accessed via end doors. The same is true of solid-sided road vehicles. Indeed, end-loading has some positive advantages. Vehicles do not have to enter the building, so no indoor space is required, and being able to run forklifts right into a vehicle is often the quickest way to process a uniform palletized load.

If you do not happen to have suitable docks (and the working areas of many buildings are located entirely on one level) it is physically possible for operatives to climb into the back of a trailer or container and to provide them with a hand pallet truck. They can then manually move pallets one at a time to the rear where they can be lifted off by forklift. It has even been known to tie a rope around each pallet and drag it to the rear, hoping it does not collapse. However, these are *not* techniques I would recommend – too many ankles have been broken in this way.

A much better solution is to use a container ramp: if there is no indoor space available it can be used in the yard (Figure 4.5). This is a portable wheeled ramp which can either be manoeuvred by hand or towed by forklift, and enables a forklift to drive up from ground level into the container. Height can be adjusted, usually by a lever-operated hydraulic pump, but an electric motor is available as an optional extra. Handrails are recommended if personnel will climb it regularly, and wheel chocks should be used for health and safety reasons. This is a good flexible option, and the solution I would recommend if permanent dock levellers are not feasible for any reason.

Figure 4.5 Container ramp in use for loading a container

SOURCE Reproduced by kind permission of Thorworld

Dock levellers

The dock leveller is one of the most useful items of warehouse equipment. Put simply, it is a device that moves up and down to overcome the problem of vehicle loading heights varying relative to the height of a dock. Many modern warehouses are built with dock levellers in mind and it is common to move into a new building without dock levellers fitted, but with pits already built in, ready for their installation. A whole bank of dock levellers is a frequent sight at major distribution centres, as in Figure 4.6.

The dock leveller consists of a substantial steel platform: the rear edge is hinged level with the warehouse floor and the front edge is raised or lowered to the level of the vehicle being loaded or unloaded. Whilst manual dock levellers are available, the majority in use are hydraulic and powered by an electric motor (Figure 4.7). At the front edge there will be a lip, often 400–500 millimetres, which can be either hinged or telescopic, and which will be extended to rest on the vehicle bed.

All dock levellers should comply with the European standard EN 1398. When specifying a dock leveller, the key features are load capacity and length. The load capacity should be governed by the maximum expected load, this being the weight of a forklift plus the weight of the heaviest pallet plus a margin to accommodate change – although, obviously, higher load capacities attract a higher price. Length is a little more complicated, as a longer dock leveller will not only be more expensive but will also occupy a larger area within the warehouse, and reduce remaining space available for handling. The main determining factor is the maximum workable

Figure 4.6 Array of dock levellers at a major distribution centre

SOURCE Reproduced by kind permission of Loading Systems

Figure 4.7 Dock leveller viewed from beneath, showing hydraulic ram

SOURCE Reproduced by kind permission of Loading Systems

Table 4.1 Maximum recommended gradients for dock levellers

Loading equipment	Maximum recommended gradient
Hand pallet truck	5%
Powered pallet truck	7%
Electric forklift	10%
Gas or diesel forklift	12%
Max permitted by EN 1398	12.5%

SOURCE Loading Systems (2018)

gradient when the ramp is raised or lowered. Trailer deck heights can vary from 0.6 to 1.6 metres but it would be rare for any facility to need to handle the full range. The maximum recommended gradients for dock levellers depend on the type of loading equipment used (Table 4.1).

Most dock levellers will be equipped with an emergency stop device in case of mechanical failure but this is worth checking, and a useful optional extra is a sensor to prevent the ramp being moved unless the door is open. Also, there should be a barrier to prevent anyone absent-mindedly walking off the end of the ramp when the door is open but no vehicle is present – this can be as simple as a chain or extendable tape. Dock lights to illuminate the inside of a vehicle on the dock can also be useful.

Whilst it is normal for a dock leveller to extend into the building it is possible to construct a dock housing externally – this will have cost implications but saves internal space.

Externally, it is a good idea to provide a recess under the dock leveller. This will allow any vehicle with a tail-lift to lower it and tuck it away under the dock so that it does not interfere with the process – such a recess can be seen at the bottom of the picture in Figure 4.7. Various forms of seal to be fitted to the outside of the building around the dock are also available, including curtains and inflatable cushions. These help to reduce draughts, and in turn reduce energy consumption, and they are most important in the case of temperature-controlled warehouses.

Sensors can be fitted to the outside of a building to aid drivers backing onto the dock, typically changing a traffic light from green to red as it reaches the required stopping point, perhaps accompanied by an audible alarm. The light system can also be integrated with ramp controls, such that the red light remains on until the dock leveller has been withdrawn from the trailer and returned to the resting position. Drivers can therefore be instructed not to drive away until they see a green light. If a UK warehouse receives deliveries from left-handed vehicles, care should be taken to position lights such that they do not confuse their drivers – and vice versa in Continental Europe. Rubber bumper pads to protect the building from damage from reversing vehicles are a good idea in all cases.

Finally, it is normal to load and unload trailers with the artic unit still attached. However, this is not universal and if it is intended to handle unattached trailers I would recommend obtaining trailer supports as an additional safety precaution to prevent collapse of trailer legs when the rear part has been offloaded but the front part has not, leading to a potentially dangerous imbalance.

Scissor lifts or lifting platforms

One application for which a dock leveller is not suitable is the loading of a double-deck trailer – these, as the name implies, have an upper and lower deck, and are used to maximize capacity when loading items such as roll cages which are not stackable. Unless the vehicle has its own tail-lift the best solution will be to use a lifting platform, commonly known as a scissor lift (Figure 4.8). Other mechanisms are sometimes used. Scissor lifts can be installed in a variety of sizes and they raise a level

Figure 4.8 Scissor lift or lifting platform

SOURCE Reproduced by kind permission of Armo (UK)

platform to the required height. Roll cages can be rolled on or off between the vehicle and the platform and it is then returned to ground level. They can be used for conventional loading of, say, pallets using a powered pallet truck, but the need to raise and lower the platform slows down the operation considerably.

Lifting platforms should conform to BS EN 1570 and should be enclosed, or at least have handrails to avoid the danger of people or goods falling off the side when raised.

Smaller lifting platforms also have a role to play. Bay lifts can be used to raise and lower items to and from a fixed bay, usually only a distance of less than 1.5 metres. Goods lifts are also used to supply mezzanine floors, usually a much smaller platform with capacity for just one or two pallets.

What sort of forklifts do I need?

Pallet trucks

I have already mentioned hand pallet trucks and pallet stackers – the former are always useful and it is handy to have one or two around even in a large warehouse. If a pallet needs moving a few metres it is easier to use than having to find the keys and start up a forklift – and of course not everyone is trained to drive the forklift. They are available with lift capacity up to 2.5 tonnes, and have a service brake. Linde offer a 500 kilogramme capacity CiTi truck, which is effectively a hand pallet truck with electric motor and large wheel so that it can be used on uneven surfaces. This would be useful if outdoor use is required, or if there is a fixed obstacle such as a track way for a sliding door that needs to be crossed regularly.

To use a hand pallet truck for an entire shift would be tiring in the extreme, so to move larger numbers of pallets on floor level the best solution is an electric pallet truck. These are cheaper than forklifts, both to buy and operate; require a much simpler level of driver training; and are more manoeuvrable in tight spaces. Their main disadvantage is that they have no lifting capacity above a few centimetres. They are again available with a range of load capacities, typically from 800 kilogrammes to 3 tonnes.

There are three basic configurations:

1. Pedestrian-operated from ground level (Figure 4.9).
2. With a foldable platform on which the driver stands.
3. With a partially enclosed operator's compartment.

The choice will involve a trade-off between cost and ease of use. If you have a small warehouse that receives one container per day so the pallet truck is only to be used for part of a shift and to move pallets small distances onto the dock then a pedestrian version may be the best option. However, if you have a large warehouse and the unit is to be used over the whole shift to move pallets over long distances then the improved productivity from a ride-on version will more than justify the additional costs.

Most battery packs will have enough capacity to last for a standard shift, and will usually be designed to be easily removable. This allows one battery pack to be in use whilst a spare pack is on charge: relying on a single battery pack risks an embarrassing situation when someone has forgotten to put it on charge; it does happen! Some chargers, especially on smaller pallet trucks, are fitted with a standard 2-pin or 3-pin plug, permitting the use of a conventional electric socket, which is a convenient option.

> Another useful feature to look for is an angled operator compartment to save the driver having to crane their neck right around when reversing.

Figure 4.9 Pedestrian-controlled powered pallet truck

SOURCE Reproduced by kind permission of Linde MH

Electric pallet stackers

These are superficially similar to pallet trucks, the main difference being that the stacker has a lifting capability (Figure 4.10). The main shortcoming is that they can only be used for open boarded pallets (see above), and they are awkward to use due to the legs protruding forwards – to place a pallet on a racking beam the legs will need to be aligned with the entry points of the pallet below before the stacker can move forwards. Whilst a counterbalance stacker is available to overcome these problems, it is rarely seen. A forklift will be more versatile with a greater lift height and faster travel speed, and whilst the electric pallet truck is cheaper, the applications will be quite specialized and I would normally recommend use of a conventional counterbalance forklift truck instead.

Counterbalance forklifts

If the average man or woman were asked to visualize a forklift, they would probably unknowingly visualize a counterbalance. These are the workhorses of the industry, and are so called because they are built with heavy weight at the rear to prevent the truck toppling forward when lifting a load.

The best-selling versions are gas powered (Figure 4.11). A removable cylinder (or cylinders) of propane is fitted to the truck: this is usually of 18 kilogrammes size and typically will contain sufficient gas for an eight-hour shift (propane is most commonly produced as a by-product of oil refining) Most forklift drivers would say that they prefer a gas truck – in practical use they tend to be more responsive, have faster travel speed, especially on a gradient, and faster lift speed, and therefore greater productivity.

Whilst exhaust gasses are not particularly noxious, they can taint certain foodstuffs, and gas trucks should therefore not be used in places like bakeries. Gas cylinders running out at an inconvenient time and place can also be highly frustrating.

Figure 4.10 Pallet stackers in operation

SOURCE Reproduced by kind permission of Linde MH

This is perhaps the time to mention that gas cylinders should not be stored indoors, but upright with valve upwards in a ventilated outdoor cage – it is obviously sensible for this to be lockable to prevent theft. They should also be transported only in open vehicles – to 'chuck a few in the back of the van' is both illegal and potentially dangerous.

Electric counterbalance trucks are also widely used (Figure 4.12). The main advantage is that they are more environmentally friendly, with no emissions at the point of use, and quieter. They tend to be more compact and therefore more manoeuvrable than a gas equivalent, and they are perhaps slightly easier for an inexperienced operator to use. In terms of purchase cost, electrics are significantly more expensive to buy than gas trucks. However, they require less (and therefore cheaper) servicing and maintenance, and the cost of electricity for charging is lower, especially if charging is carried out at night on a reduced-price electricity tariff. Whilst highly variable, the average payback period is likely to be around two years (excluding any productivity differences). A notable disadvantage is that the battery pack is heavy, and if it needs to be interchanged to allow charging this will probably require the use of lifting equipment such as another forklift.

Figure 4.11 Gas counterbalance forklift

SOURCE Reproduced by kind permission of Linde MH

Figure 4.12 Electric counterbalance forklift

SOURCE Reproduced by kind permission of Linde MH

For very heavy-duty applications, a diesel truck is recommended (Figure 4.13). The best-selling diesel forklifts have a 10 tonne lift capacity, but up to 18 tonnes is available (compared with a maximum of 5 tonnes for an electric). These are designed for outdoor use, and should not be used for significant lengths of time indoors due to the health risks (not to mention unpleasant working conditions) from particulates in diesel fumes. They tend to be most used in port environments, and in heavy industries such as steel.

When selecting a counterbalance truck, the following should be taken into account:

- Gas, diesel or electric, as discussed.
- Load capacity. It should be noted that this falls off with height and with load centres further from the mast. According to the company's data sheet, the Linde E20L has a 2,000 kilogramme capacity up to 4 metres lift height with a 500 millimetre load centre, but this falls to 750 kilogrammes at 7 metres, and to less than 500 kilogrammes at that height with a 1,000 millimetre load centre. Manufacturers will provide details on request, and this should always be taken into account.

Figure 4.13 Heavy-duty diesel-powered forklift with specialized clamps for handling stacks of concrete slabs

SOURCE Reproduced by kind permission of Linde MH

- Cab size. If you are intending to operate the forklift inside containers, a reduced height cab will be necessary for it to be able to do so. However, some, notably those in the beverage industry, prefer a high cab so that the driver can see over the load even if it is stacked high on the forks. The choice will depend on the intended application.
- Side shift. This enables the driver to move the load a few inches sideways. Even the best driver will not align a pallet perfectly every time and without a side shift will need to reverse and come forwards again on a slightly different line. Omitting a side shift from a specification can save money but it will not be worth it.
- Driver comfort. An eight-hour shift is a long time to be seated, even with breaks, and an uncomfortable seat will distract a driver. Good visibility of gauges and warning lights is also desirable and some manufacturers now offer a head up display to ensure that they are within a driver's vision. Optional extras include a revolving workstation to aid reversing, heated cabs for outside use, and even a radio (these are rarely seen in the UK but are almost standard equipment in Sweden).
- Specialist requirements, such as heat resistance for use in environments such as foundries; or cold resistance with, for example, protection against condensation and icing on electrical components, and special hydraulic fluids, for use in cold stores.
- Ability of the supplier to provide a rapid response in the event of breakdown, and the availability of replacement back-up units in case forklifts are out of service awaiting repair
- And, of course, there is the usual judgement decision regarding the overall quality of the equipment, and the price.

Reach trucks

Counterbalance trucks require a very wide aisle width, typically of approximately 4 metres. There are rare circumstances in which minimal storage is necessary, perhaps because most material is cross-docked, and a single run of racking against the wall will suffice. In general, however, installing racking in this configuration would give very poor space utilization. You will therefore need a reach truck (Figure 4.14).

The main difference is that to place a pallet using a counterbalance, it is necessary to move the whole truck forwards. However, with a conventional reach truck the driver manoeuvres the pallet in line with the pallet space, and only the mast and forks then move forwards. A recent innovation, used for example in the Linde X series, is that the mast remains fixed and only the forks move forward.

The main features to look for in a reach truck are load capacity – typically between 1 and 2.5 tonnes; and lift height, for which options vary between 4.5 and 12.5 metres. There is usually no need to add a margin for flexibility in the case of

Figure 4.14 Reach truck

SOURCE Reproduced by kind permission of Linde MH

the latter – if the racking is to the maximum height permitted by the height of the building there is absolutely no benefit in paying extra money for a higher mast. Manoeuvrability is also important, as are driver visibility, and driver comfort – some are designed to reduce the amount of vibration felt by the operator, which will reduce fatigue over a shift. I have never seen a non-electric reach truck, and would recommend a spare battery pack to allow one to be charged whilst the other is in use – lift out or roller options are usually available.

If using drive-in racking, a conventional reach truck with an overall width of maybe 1.27 metres will not be suitable. Specialized equipment with a narrower chassis width, perhaps 1.14 metres, will be required.

Finally, most reach trucks are designed for indoor use only. However, some, such as the Linde G Series or the Jungheinrich ETV C16 and C20, are also configured for outdoor use. They can therefore be used for loading and unloading in the yard, although they will not be suitable for use inside a container.

If you have only a small operation, and want to avoid the expense of more than one forklift, then this could be the best answer.

Narrow and very narrow aisle equipment

As stated earlier, definitions of these terms vary, and some would consider reach trucks to be narrow aisle equipment. However, forklifts are available that permit working in much more confined aisles (Figure 4.15).

One type is the turret truck. This superficially resembles a reach truck, but the forks can be rotated and moved backwards and forwards in relation to the mast. To place a pallet, the driver would pick it up in the conventional manner, then it would be rotated by the forks through 90 degrees, the mechanism at the same time drawing it sideways to prevent contact with the racking. When aligned, the forks move

Figure 4.15 A combination truck, which can be used to place pallets in narrow aisle racking or as a high-level order picker

SOURCE Reproduced by kind permission of Linde MH

forward as a unit to place the pallet. The mast remains stationary throughout the process. According to its datasheet, the Linde A Series can operate in aisles as narrow as 1,650 or 1,740 millimetres, dependent on the exact model.

Man-up or man-down varieties are available, such that the driver either remains at floor level or is raised up with the forks. In the case of the former a camera option to aid the driver's visibility is a good idea.

Neither turret trucks nor high-level order pickers (see below) are cheap, and a combination truck is designed to fulfil both functions.

An alternative is the articulated truck: Aisle Master, Flexi and Bendi are well-known brands. In this case the front part of the truck, including the mast, is on its own separate chassis with its own wheels. It can be steered separately from the main body of the truck, which allows operation in a narrow aisle – as low as 1.82 metre in the case of the Bendi B40VAC, for example (Bendi, 2016).

> My personal experience is that handling can be difficult, and therefore slow, if this type of equipment is pushed to the very limit of its capability, and a slightly wider aisle width may well pay dividends.

Another option is the Translift Spacemate. This is an attachment that can be fitted to a conventional forklift and operated using the side shift lever to enable it to work in a narrow aisle, and later removed to enable the forklift to be used conventionally. The Spacemate was introduced in 2016 and it will be interesting to see whether it achieves widespread use.

For operation in very narrow aisles, highly specialized equipment is required. An example is the EK range manufactured by Magaziner in Germany, which can operate at heights up to 15 metres and aisle widths as low as 1.4 metres (Magaziner, 2018). VNA trucks may be guided by wires or sensors in the floor, and may even be fully automated. I would recommend anyone contemplating this option to take specialist advice before taking any action.

Order pickers

In some warehouses orders are picked in full pallet quantities, or other heavy unit loads such as 200 litre drums. In these circumstances a forklift is required for picking.

At the other extreme, especially in a small warehouse, orders will be small, all items can be reached from ground level, and no heavy lifting is required. In such a case it may be simplest to use a trolley pushed by hand, as travel distances are too small to gain any benefit from mechanized equipment.

However, in a great many cases the requirement falls between the two. Goods are received in pallet quantities, but orders are called off in small packages – such as cartons, or plastic bags containing garments or magazines – or as individual pieces. An order picker will usually be the best solution in such cases.

These come in three main types – low level, medium level and high level.

Low-level pickers are designed for picking by an operator at ground level (Figure 4.16) – perhaps from shelves or the first two levels of pallets if these are low enough to reach. The cab does not therefore rise – it is best if the cab is open on both sides to allow exit to either direction. In front will be a pair of forks on which the operator will typically have placed a pallet, and onto which packages will be placed as they are picked. The forks will probably be able to lift the pallet up to one metre or so to save the operator from bending down, with long-term risk of back injury.

A mid-level picker will have a rising cab and forks. The operator will rise up to the appropriate level, pick the item, and be able to lean forward to place it on the pallet. Typically, a mid-level picker will have a maximum fork height of 1.5 to 2.5 metres.

Figure 4.16 Low-level order picker

SOURCE Reproduced by kind permission of Linde MH

Medium height shelving can of course be accessed using a ladder (obviously it should be of a modern type with handrails and a platform at the top to ensure safety). This is much less expensive, but moving the ladder and climbing up to, say, 3 metres takes time and slows picking rates. An eight-hour shift going up and down such a ladder will also be physically exhausting for operatives, and they are therefore only suitable for occasional use.

High-level pickers are similar in principle, but the cab and forks can be raised to a much higher level (Figure 4.17) – up to 11 metres in some models. Barriers at the cab entrances will be needed for safety reasons. If frequent picking at the highest levels is required then a heavy-duty model is recommended, as this will have a fast lift speed to maintain productivity. If most picking will be at a low and medium level with only occasional work at a high level a lower specification may be adequate, and will be cheaper.

Figure 4.17 High-level order picker

SOURCE Reproduced by kind permission of Linde MH

Tow tractors

If it is necessary to move goods over a long distance on a regular basis, the best answer may be to use a tow tractor (Figure 4.18) hauling a series of trailers. These can often be seen at airports carrying suitcases to and from aircraft. Other uses may include hauling components from a warehouse to a manufacturing plant, finished goods from the end of the production line to a storage location, or moving goods from one warehouse to another in readiness for combined dispatch. Various models are marketed with towing capacities from 3 to 25 tonnes, and control may be pedestrian, ride-on, or from a fully enclosed cab.

Cheap fuel for forklifts

I have said that there are three sources of power for forklifts and other materials handling equipment: electricity, gas and diesel. There are ways of reducing the cost of each of these.

In many warehouses, the policy for charging batteries tends to be 'Plug it in whenever you are not using it'. This has operational benefits in ensuring that the battery

Figure 4.18 Tow tractor and trailer being loaded with air cargo

SOURCE Reproduced by kind permission of Linde MH

is always topped up, but it actually reduces battery life – it is recommended that traditional lead acid batteries are allowed to run down to less than 20 per cent capacity before recharging. Also, most electricity providers offer day/night tariffs, under which electricity used at night is charged at a much lower rate than that used during the day. It is relatively simple to put a timer on the charger supply (with an override to be used if necessary) to take advantage of this.

Solar panels or other sources of renewable energy can of course be installed. The rate of return on investment may not be high, but the advantages to the environment are well known. Some may feel that these benefits are the overriding factor and use renewable energy, which is free at the point of use, for recharging.

Diesel fuel used in road vehicles attracts a duty rate of 57.95 pence per litre (in the UK as at 2018). However, vehicles that are never used on the road, such as agricultural tractors, can use red or rebated diesel, which attracts duty of just 11.14 pence per litre (HMRC, 2016). It is called red diesel in the UK as it is marked with a red dye to enable it to be identified – similar fuels in other countries may have yellow, blue or green dye.

Red diesel can be bought online in 200–250 litre drums, and this fuel can be used in diesel-powered forklifts that are used only on private property. If you take this option you will need a pump to transfer the diesel to the forklift, and to protect against spillages entering drains: special anti-spill pallets are available for this purpose. It should be noted that penalties for using red diesel on the road can be severe and include confiscation of the vehicle.

Gas for forklifts can also be bought in bulk. If usage is high enough a supplier such as Flogas will be prepared to install a bulk tank. Terms and conditions will obviously apply in each case, but usually the gas supplier will bear the capital cost of installation and then charge a rental for the tank and equipment. The user will need to provide an electrical supply. Bulk deliveries of gas are then made periodically and forklifts can be refuelled from the tank. A site survey will be required before installation as bulk gas tanks are not permitted in some locations, such as within a certain distance of a railway platform.

Savings by this method will vary according to the amount of gas used, and it will be viable only if this exceeds about 5 tonnes per year: which means that it will not be viable for a single forklift operated only on one shift. If you are operating several gas forklifts, however, this option will be well worth investigating.

Recent technological developments

I mentioned above that lead acid batteries should be permitted to run down to 20 per cent capacity before recharging. This does not apply to modern lithium batteries. These also recharge more quickly (typically two hours rather than eight hours), are lighter, and do not need topping up with distilled water. At present they are about 40 per cent more expensive but this margin will reduce, and lithium batteries are likely to become the future standard.

Forklifts have for many years been fitted with audible reversing alarms. However, in a busy warehouse with many forklifts operating and reversing frequently it is easy to become immune to the alarms and ignore them – I know I have been guilty of this. One answer is to fit a light to the forklift, which projects onto the ground in the direction of travel to warn of its approach – this can either be a blue spot or even a triangle with a picture of a forklift in the middle. These are now in widespread use and are a genuine safety improvement.

The traditional key to start up a forklift is being replaced. Options available include a keypad, into which a driver enters a PIN code, or a card reader against which the driver swipes an identity card. Advantages include not only alleviating problems with lost keys, but also being able to identify who has been driving a particular forklift at a particular time, and restricting access to authorized personnel only.

This can be interfaced with forklift fleet management software. Impact sensors and cameras can be fitted to detect and record collisions. Rather than damage mysteriously appearing and no one being held accountable, it is now possible to determine when and how a collision occurred and who was driving at the time. Appropriate action can be taken if necessary but the result is more likely to be a reduction in damage. Operator productivity can be measured and abuses such as over-long breaks identified. Fleet optimization can also be improved – are there slack times when the equipment can be better utilized, or reach trucks being used to perform the functions of a counterbalance, and is battery charging being properly managed? Used properly, such systems can be a valuable management tool.

Finally, it is now possible to buy a programmable pallet stacker (Figure 4.19). It uses SIM card technology and can be programmed using a mobile phone to visit certain locations at certain times. The truck has sensors to avoid personnel and other obstructions. The technology is in its infancy but will, I am sure, be developed. The time will come when it will be interfaced with a warehouse management system (WMS). When an order is received, the WMS will select stock to fulfil that order and direct the pallet stacker to those locations to retrieve that stock. I personally do not believe that we will see the end of the warehouse operative, but this type of technology could result in a substantial reduction in their numbers.

Figure 4.19 Robotic pallet stacker

SOURCE Reproduced by kind permission of Linde MH

Figure 4.20 Road tanker making bulk delivery of forklift gas

SOURCE Reproduced by kind permission of Flogas

Figure 4.21 Portable telescopic gravity roller conveyor in use for vehicle unloading

SOURCE Reproduced by kind permission of Owens

Conveyor systems

There are a wide variety of conveyor systems available for use in warehouses, from short portable units to complex automated systems.

Probably the commonest use in the warehouse environment is to load or unload a vehicle with the aid of a portable gravity roller conveyor. These are wheeled to allow easy movement; are often telescopic, meaning that they can be extended in length (Figure 4.21); and may be flexible to enable them to operate around curves. Typically, one end of the conveyor will be extended into the vehicle, and one or two people will load items onto the conveyor. These will then be fed down by gravity and a team waiting on the dock will sort the items and perhaps palletize them.

Powered versions are also available, with either rollers or belt, and these are particularly useful for raising cartons onto a vehicle where no dock leveller is installed.

Longer, permanently installed conveyors can be used to move items over greater distances, often between a picking area and a packing or dispatch area. As an example, a well-known fashion retailer used a series of picking silos, where items were picked at one of several levels of mezzanine floor. The items for each store from each level were placed in plastic tote boxes. When each was full or the picking on that level was complete, it was sealed and labelled and placed on a conveyor that ran across the end of each aisle on each level. These were then transported down to the dispatch area where the totes for each store were consolidated throughout the day ready for night-time dispatch.

Most conveyor systems allow items to stack against each other at the end of the conveyor awaiting removal. However, where this would create a risk of damage sensors can be installed to prevent contact between items. Where it is necessary to move items up or down in a small area, for example to and from a mezzanine floor, spiral conveyors are recommended, which are shaped like a spiral staircase.

Whilst most conveyor systems are used for smaller items, heavy-duty versions are available for pallet handling, able to handle pallets of 2 tonnes or more. On a related point, the price of conveyor systems is affected greatly by their load capacity. If your requirement is mainly for articles of a particular size, but occasionally larger items, then it will probably be more cost-effective to specify a lighter capacity conveyor system and to use a manual solution for the few large units.

Finally, conveyors can be incorporated into automated systems for sortation, and this is the most common application for newly installed systems at present. Barcode readers are often installed alongside the conveyor, and the system is programmed to

Figure 4.22 Swivel roller sorter diverting a parcel to the side conveyor on the left

SOURCE Reproduced by kind permission of Axiom GB Ltd

divert items off the main conveyor run onto a chute or side conveyor at the appropriate point, for example to segregate parcels according to the vehicle onto which they will be loaded (Figure 4.22). This can be done by tilting, pushing, using angled swivel rollers, or using a cross belt. The most complex systems can sort tens of thousands of items per hour.

Conclusion

In this chapter I have described the various types of equipment available for use in warehouses of all sizes. This has included fixed installations such as dock levellers, mobile equipment such as forklifts, and conveyors which can fit into either category; and items varying in price from tens of pounds, euros or dollars to tens of thousands.

In the next chapter I will discuss how this equipment can be used to provide a first-class warehouse service.

Warehouse operations 05

It is too often forgotten that the key people in a warehouse are the operatives. In the final analysis, the key factor in whether a warehouse runs well or runs badly is whether the operatives do their jobs right – for example picking the correct items at a productive rate without damage. Management of course has an important role in ensuring that people have the right resources, are fully trained and well-motivated, but it is the operative on whom all ultimately depends.

In previous chapters I have described the various options for the physical equipment available for warehouses (racking, forklifts, etc) to ensure that the best facilities are available for operatives. I will now discuss how operatives can best use those facilities.

Does the process start with receiving?

If someone were about to become a parent for the first time, would they consider the process starting with the baby's birth? The only sensible answer to this question is 'No' – and the same principle applies to logistics.

The process should begin at the purchasing planning stage. The major retailers have very strict requirements for their suppliers, and have sufficient purchasing power to enforce these. You may have less purchasing power, and would not be able to make such precise demands, but there is nothing to stop you making reasonable requests, and sometimes it will be a case of 'You only needed to ask'. Your 3PL, if any, might also have a useful contribution to make in specifying requirements that will make it easier for them to provide you with a first-class service.

First, there should if possible be a three-way discussion between the logistics, sales and purchasing departments about pack quantities and dimensions, so that these are consistent throughout the supply chain. Some companies, for example, import goods in pack quantities of 12 but sell in pack quantities of 10, with the result that their 3PLs make a great deal of money transhipping from one size of box to another. If purchasing were able to negotiate a pack size of 10, then the cost of labour and of boxes would be saved. Similarly, presentation in a retail pack, labelling and barcoding can be carried out much more cheaply in the Far East than in Western Europe. If you are purchasing in significant quantities such improvements should be negotiable.

Second, the way in which large quantities of cartons are presented should be agreed. They will be easiest to handle if they are palletized, and if purchased domestically or within Western Europe this would be the norm. However, pallets take up shipping space (up to 10 per cent of the volume of a container), some countries such as Australia have strict regulations on import of wood, and the quality of pallets provided may not be up to the standard you might wish. For these reasons pallets are the exception rather than the rule with deep-sea containerized imports.

One alternative is the use of a slip-sheet (Figure 5.1). Cases are stacked and wrapped, but rather than being placed on a pallet they are placed on a flat sheet (normally plastic or cardboard) called a slip-sheet. On arrival they are unloaded by a forklift with a specialized push–pull attachment, and can then be transferred as a unit to a pallet. This system is widely used for New World wines from Australia, Argentina, South Africa etc.

Figure 5.1 A slip-sheet attachment in use, transferring a stack of boxes onto a standard pallet

SOURCE Image courtesy of B&B Attachments Ltd (www.bandbattachments.com)

If cartons are to be individually stacked it is not unreasonable to expect them to be stacked neatly, and different items packed together. I have seen examples of cartons that have clearly been thrown into the container so they are at all angles in a heap, and of containers with 2,400 cartons across 12 different part numbers stacked entirely at random. Both should be unacceptable. Stacking should be neat and regular, and boxes of each item should be stowed together.

Carton sizes should also be agreed with pallet size in mind. Cartons of 200 x 300 millimetres will stack very well on standard 1,200 x 1,000 or 1,200 x 800 millimetre pallets; cartons of 337 x 173 millimetres will create a problem.

It should be stipulated that all goods should arrive with an advice note. Paper is still the most common format, but it can be sent electronically, and paperless receiving is likely to become future standard practice. This should include details of the supplier, part numbers, descriptions, quantities, types and numbers of packages, purchase order numbers, and any other relevant information you may specify. Goods should also be clearly labelled. These are perfectly reasonable stipulations, even if you are just one of many customers of a supplier. For your part, you need to give clear instructions, for example relating to delivery address: attempted deliveries to accounts or purchasing departments are more common than you might think.

Equally importantly, full information regarding the products should be provided in advance. This should include:

- stock-keeping unit (SKU) number;
- description;
- customs commodity code if relevant (see later chapter);
- special requirements, for example regarding hazardous goods and weights, dimensions and quantities per pallet, outer carton and inner carton;
- weights, dimensions and quantities per pallet, outer carton and inner carton.

The information can then be pre-entered onto the warehouse management system (WMS) – the alternative is that measurements of goods are taken on arrival, new items will need to be set up on the WMS, and in practice the goods will have to be left cluttering up the receiving area whilst all this happens.

The warehouse supervisor will also be able to prepare for any special requirements, which may not be obvious without notification. Examples from my personal experience include tea, which one may think of as light, but was actually delivered in 60 kilogramme sacks, so warehouse staff required specialized manual handling training; and drums of washing-up liquid concentrate, which is classified as hazardous, even though washing-up liquid itself is not. Specialized equipment may also be needed, such as forklift attachments, for example:

- barrel clamps for drums of liquid;
- flat-sided clamps for white goods or bales of fabric;
- booms for rolls of carpet.

Finally, one mistake to avoid. If you are receiving seafreight containers, they will probably have a hardened steel bolt seal. Before it arrives, invest in a decent pair of long-handled bolt cutters. Sawing through with a junior hacksaw is difficult, and having to run out to the local DIY store to buy a junior hacksaw when the first container arrives is simply embarrassing.

Slot times and pre-advice

It is also quite reasonable to expect at least large deliveries to be allocated a specific time for their delivery – known as a slot time – and to expect the delivery company to adhere to this. For full load artic and container deliveries this is standard practice, and hauliers will expect this, as will those delivering numerous pallets.

Single pallet deliveries will generally attract a surcharge for a timed delivery, £15 being a typical figure in 2018, which you may or may not consider worth paying, as your supplier will probably want to pass on the cost. Deliveries of small numbers of parcels cannot be expected to be timed – the driver may well have over 100 deliveries to make and would have to do so in the most efficient order.

The warehouse supervisor will need to lay down guidelines for slot bookings according to the resources available. There may, for example, be four dock levellers, but only enough operatives and/or enough equipment to handle two containers of unpalletized goods at any one time. Sufficient time for each job must also be allocated – a container of palletized goods might be discharged in 30 minutes, but one of individual packages might take three hours. It is common to offload incoming vehicles in the morning and to load outbound in the afternoon and evening, but this is not universal, and some 24-hour operations both ship and receive throughout the day and night.

> It is advisable not to book every slot too far in advance, so that genuinely urgent requirements can be accommodated at short notice.

Many warehouse management systems have a slot time function. However, a spreadsheet can be used effectively (Table 5.1), and sometimes the old ideas are the best – a big white board on the wall can be seen by everybody.

Table 5.1 A fictional example of a slot time schedule

Slot time bookings   Day: Monday   Date: 8.7.19

Time	Dock 1						Dock 2						Drive-in Dock					
	Haulier	In/Out	Supplier/customer	Quantity		Notes	Haulier	In/Out	Supplier/customer	Quantity		Notes	Haulier	In/Out	Supplier/customer	Quantity		Notes
				Plts	Ctns					Plts	Ctns					Plts	Ctns	
6.00	Own artic	In	CHEF	500		Empty pallets	K Line	In	Haidong		1800	KKFU1177998	Speedy Courier M/bike	Out	Sales Director home		1	Urgent
6.30	Own artic	Out	Jacksons Wakefield	26		Double slot	K Line	In	Haidong		1800	KKFU1177998						
7.00	Own artic	Out	Jacksons Wakefield	26		Double slot	K Line	In	Haidong		1800	KKFU1177998	Own van	Out	Potters Bar Garden Centre		74	
7.30							K Line	In	Haidong		1800	KKFU1177998						
8.00	MK Haulage	In	RSJ Milton Keynes	6			K Line	In	Haidong		1800	KKFU1177998						
8.30							K Line	In	Haidong		1800	KKFU1177998						
9.00													M&P	In	Weston Garden Centre	1		Return
9.30																		
10.00	Maersk	In	Weifang		2400	MSKU07002682	K Line	In	Haidong		1972	KKFU7769021						
10.30	Maersk	In	Weifang		2400	MSKU07002682	K Line	In	Haidong		1972	KKFU7769021						
11.00	Maersk	In	Weifang		2400	MSKU07002682	K Line	In	Haidong		1972	KKFU7769021						
11.30	Maersk	In	Weifang		2400	MSKU07002682	K Line	In	Haidong		1972	KKFU7769021						
12.00	Maersk	In	Weifang		2400	MSKU07002682	K Line	In	Haidong		1972	KKFU7769021						
12.30	Maersk	In	Weifang		2400	MSKU07002682	K Line	In	Haidong		1972	KKFU7769021						
13.00																		
13.30																		
14.00	Garston	In	RSJ Warrington	26														
14.30																		
15.00							Davies Turner	Out	Export Groupage	14	32							
15.30																		
16.00																		
16.30	NBC	Out	Bailey Glasgow	26														
17.00							NBC	Out	Bailey Newcastle	26								
17.30	NBC	Out	Perkins	18														
18.00	NBC	Out	Hopgood	22			Own artic	Out	Jacksons Dudley	26								
18.30	NBC	Out	Wade	8	16	18 tonner booked												
19.00							Own artic	Out	Donalds	26								
19.30													DHL	Out	Next day parcels		TBA	
20.00	Depot closes																	

There should also be a pre-advice of what is expected to be received – this will commonly take the form of an advanced shipping notice (ASN) from the supplier, and should include similar details to the advice note. This will enable the warehouse supervisor to plan for sufficient labour to be available to process the consignment, and for a pre-entry on the WMS to save time on arrival.

The company's procedures should include a method of ensuring that material has genuinely been ordered. In some cases, especially in smaller companies, goods are only ordered when needed and this will simply entail checking on the system that the purchase order is valid and the items and quantities are correct. If, however, the company is operating a blanket order system, under which a purchase order covers, say, six months' supply to be called off over that period, or even states only a price and no quantity, communication is required to ensure that deliveries are made only in line with demand. It is to the supplier's advantage to ship as early as possible, to avoid storage costs of their own and to improve cash flow by earlier billing, so deliveries need to be managed to avoid abuse.

Finally, having stipulated that slot times must be adhered to, there is a responsibility on your part to unload vehicles promptly. Failure to do so will often result in your being invoiced for demurrage (a charge levied by hauliers to offset additional costs incurred due to vehicles being unproductive whilst delayed), and will have knock-on effects such as drivers running out of hours or collections not being made at the haulier's other customers.

As an extreme example, one major distribution centre treated this matter with disdain, with the result that at least one nearby organization specifically wrote into their contract with hauliers that vehicles booked for collections must not have been rostered to deliver at that location within the previous 24 hours.

Yard management

This is an often-neglected area of warehouse management. Many drivers will all be able to tell tales of wandering around a building trying to find out whom they should inform of their arrival, and of seeing someone standing on a dock leveller waving franticly and shouting, 'Oi, driver. Over 'ere, mate.'

The process will be easier if there is a manned security barrier at which a driver can stop, report and be given instructions. Some large facilities have excellent infrastructure, for example with large screens onto which messages are displayed. However, this will not be the case with smaller premises. When a driver arrives in the yard it should be clear where the vehicle should be parked, and there should be a sign over the appropriate door (or intercom connection) saying 'Drivers Report Here'. Docks should be numbered so that they can be given clear instruction – 'Please wait in the parking bay until the artic on dock 3 moves off, which will be in about 15 minutes, then pull onto that dock' or whatever. This also ensures the right person has the paperwork and can do whatever is necessary, such as allocating a team to unload.

If loads are received from overseas, drivers may not speak English. It may help communication, and certainly make a driver feel more welcome, if someone speaks at least a few phrases of their language, and to keep a list of commonly used phrases that can be used. Free online translation websites may not recognize specialist logistical meanings of some words.

For example, the metal plate on the back of an artic unit, into which the kingpin of the trailer fits, is called a fifth wheel. In French it is *une sellette* (Gefco, 1996), which translates literally as a small turntable. Unless they happen to be familiar with the English term, to talk to a French driver about *une cinquième roue* would not make much sense.

Drivers who are on site for any length of time should be given access to toilets, and it should be noted that some drivers are women. For a female driver to arrive with a night trunker and be informed that there is a lock on the women's toilet and that only the women in the office know the combination would be incredibly inconvenient. Indeed, the then UK Transport Minister Claire Perry cited the need for women to 'relieve themselves behind a bush' as one of the reasons that women make up such a small proportion of truck drivers (Perry, 2016). It is also good practice to allow drivers access to a vending machine or a kettle and some tea bags, milk and sugar – this will certainly help to gain their cooperation if it is needed.

Finally, do not forget health and safety in the yard. If possible, private cars should not be permitted, although in some cases parking space is at a premium and this is not feasible. Walkways should be clearly marked and where possible protected by barriers, and of course fire exits must not be obstructed.

Drop trailer operations

It is usual for vehicles to arrive, be loaded or unloaded in the presence of the driver, and depart. However, some organizations use what are called drop trailers. This means that an artic will arrive, and the driver will uncouple the trailer, to be unloaded at some later time in their absence. They will then couple their tractor unit up to a different trailer before driving away. Sometimes the trailers are left on a bay, and

Figure 5.2 Tugmaster showing a lifting fifth wheel

SOURCE Reproduced by kind permission of Sotrex.com

in larger operations there are one or more shunting vehicles on site to move trailers backwards and forwards between loading areas and a trailer park. Specialized purpose-designed vehicles called tugmasters are available, with a small cab, excellent all-round visibility, powerful engine with low gears, and a fifth wheel on a liftable boom, which saves having to wind trailer legs up and down (Figure 5.2). These are most commonly seen in ports hauling unaccompanied trailers on and off ferries, but can also be found in some very large drop trailer operations.

The main advantage of drop trailer operations is that drivers and artic units are not detained awaiting unloading – thereby avoiding unproductive costs. On a small scale a trailer may be left on a bay for loading throughout the day and an evening dispatch; on a larger scale 50 trailers may be trunked in and out during the night for processing during the day, or vice versa. There is obviously a cost in that extra trailers are required, but in a well-managed operation they can be more than offset.

Receiving and checking

I have talked in the previous chapter about the various types of equipment available for offloading, and there is little to add. However, I will make some general points.

First, double handling should be avoided whenever possible. If a mixed load of cartons needs to be palletized, it is not uncommon to see them all being placed onto pallets, and these being pulled off the vehicle by a powered pallet truck into a holding area. There they are re-palletized into separate pallets for each SKU. It is more efficient if they have arrived with cartons of each SKU stacked separately, so they can be placed directly onto the right pallets; or placed on a conveyor so that the operatives at the other end can have an array of pallets waiting.

The stock should be offloaded into a checking area, and an immediate 'major pack' quantity count made, so that any discrepancy in the number of pallets, cartons, etc, or any obvious damage, can be notified immediately to the driver. I would recommend that as a security measure you have a unique rubber stamp made up for confirming receipt. It is not unknown for 'proofs' of delivery to be signed by Michael Mouse, PG Tips or (my own favourite example) Aaron C Rescue: if the delivery note does not bear the correct stamp you will know that the proof of delivery is not genuine. The stamp should include the words 'Received Unexamined' so that discrepancies that have not been picked up immediately can be notified later.

For example, you might have ordered 144 reels of coaxial cable and 288 of three core cable, but actually received 132 and 300 respectively, which only comes to light at a later stage.

The question of thoroughness of checking is a difficult one. Certainly, high-value items should receive a thorough 100 per cent check, and it is prudent to do so with new suppliers or those who are known to be unreliable. There are two main types of check:

- check against expected;
- blind count.

For the former, an operative will have a listing of what is expected, and tick off what they find against that list. For the latter, the operative will list what has arrived, and only after completing this list will they then check against what was expected. The blind count takes more time and will therefore be more expensive to conduct. However, there is a human tendency to see what one expects to see and the blind count is therefore more likely to pick up discrepancies. Opinions will differ as to the best choice.

Whichever option is chosen, a second person checking apparent discrepancies is a good idea, and any shortage or damage found should be notified to the haulier and the supplier immediately. A record should be kept, not only to enable a credit against the supplier's invoice to be obtained, but to identify trends and investigate further if necessary: for example, there may be repeated damage on a particular item, which might indicate inadequate packaging.

In the case of damage, the record should include a photograph, and whilst most people will have a camera on their mobile phone there may be rules discouraging them from the working area. I would therefore recommend that a cheap digital camera be kept in the receiving area.

If suppliers are known to be reliable, 'good faith' receiving can save significant sums of money by avoiding checking. Under this system only a percentage of consignments are checked, and discrepancies charged pro rata, so that if 10 per cent of items are checked revealing £50 of damage, £500 will be billed to the supplier. This system is in use by some major retailers.

Checks can be eliminated altogether and costs can be shared. As an example, one major motor manufacturer had five parties involved in their deliveries to UK dealers: continental transporter company, continental port, shipping line, UK port and UK transporter company. Each was checking every car for minor dents and scratches. However, it was noticed that each of the five checks cost more to carry out than the average cost of all damages per car across the whole delivery system. A *modus vivendi*, or way of life agreement, was therefore made under which no checks were made, and all costs were shared equally between the five parties, saving not only money but a lot of time and trouble.

Finally, it is important to maintain integrity in the checking process, and ensure that no goods are removed from the checking area until the process is complete. There is a temptation to grab an urgently needed item, either to maintain production or fulfil an important sales order, with the intention of completing the formalities later – but frequently not doing so. Disciplines should be in place to prevent this, but I am afraid I have no useful advice if the person who has a habit of breaking the rules happens to be the boss.

Put-away

Before goods are put away into storage, there should be instructions laid down as to which items are held where.

The long-established advice on this matter is to conduct what is called a Pareto analysis and classify SKUs as group A, B, and C according to throughput: the number of orders which include that SKU is usually the best measure. In a typical case

20 per cent of the SKUs will account for 80 per cent of movements. These should form group A, and be placed in the locations nearest to the shipping and receiving bay. The next 30 per cent of SKUs will account for another 15 per cent of movements (95 per cent when combined with group A). These should form group B and be in the next most easily reached locations. The remaining 50 per cent of SKUs (group C) will account for only 5 per cent of throughput and should be in the least accessible locations.

In some cases, this is indeed the best advice. I knew of a confectionary company that shipped only in full pallet quantities to retailers, and this system worked very well. The one point I would make is that height of locations should be taken into account – it will take longer to place a pallet at the top of racking than to travel a short distance along the aisle and place it at ground level.

However, there are certain circumstances in which it should be overridden. For example:

- The warehouse is too full to permit it. If one tries to run a warehouse at close to 100 per cent capacity, or even in some cases at over 100 per cent, an incoming item will need to be placed in whatever location is free even if that is inappropriate. This is one of several reasons why productivity falls dramatically if one operates at above about 95 per cent capacity and this should be avoided if possible.
- Types of storage. This will usually be an obvious requirement – a small area of cantilever racking installed for long items must be used for those items as they cannot be stored elsewhere; or high value items will be kept in a separately secured area. Also, if there is pallet flow racking for, say, 25 SKUs this will in practice dictate that group A consists of 25 SKUs regardless of the optimum number suggested by the result of the Pareto analysis.
- Closeness of replenishment stock. If goods are received in pallet quantities and shipped as cartons, an efficient method (if it is possible) is to store a single pallet of each SKU at floor level for picking. It is most efficient if the reserve stock of full pallets of the appropriate SKU is stored directly above each picking pallet.
- There are a small number of items that appear in almost every order. An example of this is a toy-maker, for whom most orders require batteries. These can be classified A* and placed in very accessible locations, such as the end of each aisle.
- There are SKUs that tend to be called off together. Consider an electrical wholesaler who supplies contractors. If one particular contractor is carrying out a security installation the order will probably include access card readers and motorised locks; a satellite installation will include a satellite dish, mounting brackets and coaxial cable; a garden lighting installation items designed for outdoor use, and so forth. In these circumstances items that tend to be called off together should be stored together.

- Seasonality and promotions. There are few products that do not have greater call-off at certain times of year, and some such as Halloween goods show an extremely short but dramatic spike in demand. (Some pharmaceuticals have constant year-round usage, and washing powder has just a small downturn in November and December as supermarkets clear their shelves for Christmas products.) If, for example, your product range includes heaters and fans, these will fall into different groups at different times of year. Material for an upcoming promotion might be considered part of group A even if past sales do not justify this.
- Items purchased to order. In some environments particular items are purchased only when required, rather than being held in stock. An easily accessible location for such items is obviously a good idea, preferably one in which all items for that order can be held together. Assembling goods from stock that will also form part of that order into the same location is an option I would favour, where feasible.

There is no part of the warehouse management process for which accuracy is not of the utmost importance, but this is especially so of the put-away process. A variety of aids are available, such as barcodes, radio frequency identification (RFID) or voice. More will be said on this subject in the next chapter.

Whichever method is used, an item in the wrong place is an item lost, and searching the warehouse for a lost item occupies far more time than not losing it in the first place. All storage locations should be individually identified, and subdivided if necessary: storing different parts in the same location is best avoided. This is sometimes forgotten with abnormal items. I knew of one car parts warehouse in which most items were shelved in a professional manner, apart from drums of oil and hydraulic fluid, which were all held together in the location 'floor'. This was where most problems occurred, and it would have been far better to have bought a can of paint to mark out specific areas of the floor, each to be treated like any other location.

Stocktaking and perpetual inventory

Stocktaking is not an enjoyable responsibility. There are many benefits to maintaining accuracy in a warehouse, and one of these is that stocktaking will be required less often. Whilst there are firms that specialise in offering stocktaking services, their fees tend to be expensive.

It is usual to conduct a complete 100 per cent check of all stock at least once per year: sometimes it will be six-monthly or even quarterly. For most businesses this is an audit requirement, and auditors may wish to attend and verify that the stocktake is accurate.

The timing of the stocktake should ideally be at the quietest period of the year, when stock is at its lowest. This will often mean immediately after Christmas, and this conveniently coincides with the end of many companies' financial year. Some operations attempt to conduct a full stocktake whilst still maintaining a full service, but this rarely works well in practice. It is far better if it is possible to stop the service completely, and if this means stocktaking over a weekend or holiday period then so be it.

Preparation for the stocktake should start well in advance, and include the following:

- Announce the date several months ahead, and ensure that people required, especially key personnel, have not booked their holiday at that time. Ensure that others involved, such as auditors and your own finance department, are available on those dates to avoid the need for an inconvenient last-minute change.
- Ensure that the necessary equipment is available – for example, will you need to hire a cherry picker to be able to check the highest levels of the racking safely?
- If possible, complete all the 'When we get around to it' jobs, such as disposal of obsolete stock, before the stocktake. In reality, however, this may not be possible if the stocktake follows shortly after a busy period such as Christmas.
- Stop incoming shipments from a specified time, ensure that any shipments received before that time have been fully processed and put away, and that all warehouse management system processing is complete. Similarly, no outbound consignments should be picked until the stocktake is complete. To have to count material in the shipping and receiving area is an unnecessary complication. If receiving freight is for some reason unavoidable, it should be kept in a quarantine area until the end of the stocktake.
- Tidy up! The process will be a lot quicker and more accurate if material is neatly stacked ready for counting – this especially applies to partly picked pallets. Also check that everything is labelled as it should be. If a label has fallen off it is far better to identify the goods and re-label before the stock check takes place.
- Ensure that everyone involved is fully briefed and familiar with their duties.

The actual process of the stocktake may well be driven by the capabilities of the WMS. If it is possible to conduct the check by scanning barcodes or reading RFID tags this will aid both speed and accuracy. However, if these are not available the most common method is to generate a list of locations, perhaps for each aisle. This may or may not include the SKU that should be in that location, but should in either event include a blank space beside each location into which a quantity is entered as the goods are counted. On completion of the aisle, this is compared to a similar sheet, which has been printed with quantities. If the expected and actual quantities match, all is well. However, if there is an apparent discrepancy a supervisor should recheck the location to determine whether the error lies in the stock quantity on the WMS or in a miscount during the stocktake.

The other element of stock control is cycle checking, or perpetual inventory (PI). This involves checking a small proportion of locations each week. This can be done either by working systematically through the warehouse (aisle 1 this week, aisle 2 next week, etc) or by randomly generating locations to be checked – most WMS have a function to do this. Discrepancies are more likely to occur where there is the fastest turnover of stock, and Richards (2018) suggests checking 8 per cent of group A items, 4 per cent of group B and 2 per cent of group C on each cycle check. An alternative is to instruct operatives to count remaining stock every time they make a pick. This will, however, mean that a great deal of time is spent on stock checking which would only be justified if the accuracy of picking is very poor. I would recommend additional training or other measures to improve accuracy rather than taking this course.

If cycle checks reveal few discrepancies, their frequency can be reduced. If one is able to reach the stage where nominal differences from recorded stock are found, it may be possible to convince auditors to eliminate the annual stocktake completely, which would I am sure be a relief to everyone.

Finally, stock adjustments may be triggered during routine operations: for example, an operative may go to a location to pick an item and find it empty. If this is occurring regularly it is almost certainly a sign of an underlying problem.

Discrepancies (however they are detected) should be monitored, any recurrent problems identified, and their root causes determined. Sometimes this will be due to a misunderstanding.

For example, an operative may have an instruction to pick three bottles of wine, but actually picks three cases. Clearer instructions and/or additional training would be the answer here.

Another problem that must be firmly addressed is lack of discipline. If an operative has a pick list including items from a location at the far end of the warehouse then they need to be picked from that location (unless there are stringent manual override controls in place). There may be a temptation, if passing the same item en route, to pick from the easier location, which will save a little time in the short term but create problems when a picker later finds the nearer location empty. Unless properly addressed, such problems tend to multiply and become a recipe for chaos.

There is, of course, the threat of theft from many operations, and stocktake discrepancies can be an aid to detecting this. Some years ago, an airline conducted an exercise to examine shortages found when in-flight retail trolleys were returned to the warehouse for restocking. It was found that certain high-value items were found to be missing regularly on flights when certain cabin crews were rostered, which ultimately resulted in some prosecutions. However, the most common missing item was found to be Toblerone chocolate bars. A firm 'Stop this' message from management reduced the problem substantially.

Picking

Some would say that picking is the crucial element of warehousing, and all other activities are carried out simply to support the picking process. I would not go as far as to say that, but certainly it is the most labour intensive, one of the most difficult to get right, and one of the most difficult to automate. It is, if anything, becoming even more labour intensive, as just-in-time purchasing and online shopping reduce the size of individual picks, but increase their number. One of the reasons for complexity is that picks can be measured in different multiples for the same SKU – in some instances it will be possible for a customer to order a whole pallet, a whole layer, an outer carton, an inner carton or an individual item. This is further complicated if goods can be ordered by quantity – eg cable might be ordered either by the metre or the full reel, or hydraulic fluid may be available by the drum or by the litre. Clear differentiation, and thorough training of pickers to ensure they are familiar with the variables, are essential.

Richards (2018) provides measures for picking accuracy (see Table 5.2). To put this another way, any customers of a best in class operation who receive orders daily will have a problem with an order about once a year. This is excellent service with which any reasonable customer would be well satisfied. At the lower end, however, such customers would have a problem with an order about once a week. Few would find this acceptable and they would probably soon be seeking a new supplier, so robust processes are therefore essential. It is also worth remembering that for every short shipment, which most customers will be very quick to report, there will probably be an over-shipment, which many customers will not report at all. The cost of underperformance can far outweigh the cost of rectifying that underperformance.

Table 5.2 Measures for picking accuracy

	Best in class	Median	Major opportunity for improvement
Orders correctly picked	≥ 99.8%	99.3%	≤ 97.0%
Orders on time in full (OTIF)	≥ 99.59%	96.53%	≤ 82.8 %

The old-fashioned method was to give a picker a sheet of paper listing all the items on an order. They would then move around the warehouse, pick those items (ticking off the list as they go), take them to the shipping bay, then pick the next order, and so on. To maintain the level of concentration required to achieve 100 per cent accuracy using this method, throughout an eight-hour shift, is something very few can achieve: in reality the mind will drift to other matters, whether it be a child being bullied at school, or who will be the next to be eliminated on a TV reality show. I worked in a warehouse during my gap year in 1979, mainly in the office but sometimes as a picker. This was before computerization, and I am willing to admit that I did not achieve 100 per cent pick accuracy using this method.

A great deal can be achieved by good personnel management – training, motivating, instilling discipline and a positive quality ethos, performance monitoring, and so forth. This cannot, however, achieve perfection.

In some warehouses all orders are rechecked prior to dispatch, which is time-consuming, but can be the only way to eliminate errors in a small warehouse where the expense of investment in technology cannot be justified. However, some form of technological assistance is highly desirable, and various options for this will be discussed in the next chapter: barcodes, RFI, pick by light, etc.

I will, however, mention now one of the simpler technological devices to use, ie automated check weighing. A version of this technology is used in the self-scanning area of a supermarket – if one scans a box of six eggs but places a bottle of Champagne in the bag an automated voice will announce 'Unexpected item in the bagging area'. In the warehouse environment, the weight of each item is held in the WMS. The expected total weight for the consignment is calculated, and compared with the actual weight, a difference indicating a discrepancy. The system is not fool-proof – for example, two different SKUs may be of the same design, and therefore weight, and vary only in colour – but it does make a very useful contribution.

In the following sections I will lay out the options for picking methodologies, whichever technology is used.

Picking methodologies

In some instances, walk time can take up 50 per cent of the total time for the whole pick process. To maximize productivity, walk time should therefore be minimized, and a well thought-out put-away system (see above) will now pay dividends. Also, a good WMS will be able to generate a pick list in the optimum order, with the shortest walk distance and ending at the shipping area – if a list is in, say, numerical order of SKU number there is a danger of missing an item and having to backtrack to pick it.

With this in mind, the main options are as follows.

Individual order pick

Under this system a picker is given a single order, and picks that order, before returning to the start and picking the next. This is the easiest system from the picker's viewpoint, and as it is usually the quickest way of picking a single order it is the best method to use for an urgent order. It is also the standard method for large orders: the picker may be unable to carry more than one consignment of that size on their order picker, for example. It also avoids the risk of items getting into the wrong consignment, which might occur if the picker is working on several orders at once. The disadvantage is that there is no synergy between orders and a picker could visit the same or nearby locations several times during their shift, so that walk time is greater than for other methods. It will be especially inefficient if orders consist only of a few items each, as the picker will be returning repeatedly to the start point.

An example would be a car parts warehouse, where a dealer places an order for perhaps 100 different line items.

Cluster pick

Under this system, a picker takes several orders at once rather than a single order. As they pass each location they pick the items, and place them with the other items for the appropriate order. The advantage is that total walk time is reduced, and therefore more orders are picked in a given time, but there is a risk of an operative getting 'mixed up' so that an item is placed into the wrong order. This is a method I would expect to see used in some mail order warehouses, where orders most commonly consist of a few line items only.

Batch picking

Under this system items for a number of orders are picked, then taken to a secondary handling area. Here a secondary pick takes place as they are allocated to individual orders. In most situations, where orders are widely varied, this will be inefficient as it almost duplicates the picking workload. However, it can be beneficial where many orders are similar. One example would be a TV shopping channel, where most orders are for the item currently being broadcast. Another would be a promotion, with an offer to sell a specific mix of products at a heavily discounted price.

Zone picking

This method will usually be a version of one of the above, where picking is subdivided into zones. The pick from each zone is then combined to make up the order. This can be used simply to reduce the risk of pickers getting into each other's way, but most commonly it is driven by the geography of the warehouse. Usually the picks are separate, and combined only for dispatch. However, it is possible for the pick from one zone to be placed into a tote box or other receptacle, and for that tote box to be taken to the start point for the next zone so that picking can continue into the same tote box. For example, in the car parts warehouse mentioned earlier, there may be one area for large items such as engines and body panels, and another for small parts such as wiper blades and light bulbs. Separate zonal picks from each area could then be combined in the dispatch bay. An extreme example was a well-known fashion retailer, who kept boxed garments in a warehouse in the northern UK, and hanging garments in a warehouse in the south. Picks were made from both warehouses each day for stores across the whole country, and each night the hanging garment picks for northern stores were trunked to the northern warehouse, and combined with the boxed garment picks for deliveries to each high street location. The system worked the opposite way for boxed items, and it did work well.

Methods not mutually exclusive

It should be stressed that these options are not mutually exclusive, and it may be appropriate to use different methods at different times in the same operation. An example might be a wine importer. They might organize picks as follows:

- orders for independent off-licences, each of 50 to 100 cases – individual order pick;
- orders for mail order customers, consisting of six or 12 bottles – cluster pick;
- promotional orders, where 500 customers order the same combination of six different bottles – batch pick;
- order for off-licence which includes two display stands – combination of individual order pick and zone pick, with the two stands coming from a separate building.

Goods-to-picker

All of the above methodologies have assumed a picker-to-goods strategy. However, it is possible to take the goods to the picker. I have seen this done by using reach trucks to collect pallets from the racking to a handling area, where a picker removes one or more cartons, and the pallets are then returned to the racking. This is inefficient and I would recommend hiring a high-level order picker as an alternative.

However, it is possible to do this by an automated system. I have mentioned that I have been fortunate to visit the John Lewis facility in Milton Keynes, and this provides a very impressive example. Pickers are located at ergonomically designed workstations, with materials such as empty boxes at hand. Goods are stored in an automated warehouse, often in plastic totes. These are selected by robotic equipment in line with information provided by the WMS, and totes containing the various items required to fulfil a particular order are moved to the picker. The correct number of items are removed, this being confirmed by a real time weight check, and the tote with its remaining contents returned to storage. When the order is complete the picker places the box on a separate conveyor and the box is sealed and labelled robotically en route to the dispatch area. The advantages are clear, with completely eliminated walk time, no time wasted looking for the right SKU, reduced fatigue, etc, leading to dramatically improved productivity. Less obvious advantages include improved opportunities for disabled members of the workforce. Such automation is not cheap, and will not be viable in small operations. Richards (2018) recommends that it be considered if throughput exceeds 3,000 cartons per day.

Sequence of order picks

Part of the planning for warehouse activity should be to ensure that all orders are picked on time. The simplistic methodologies include picking the oldest order first, or the order for the most important customer first. However, these can create problems in practice – in the first scenario the oldest order may not be the most urgent, and in the second an order may never reach the top of the list and therefore never get shipped.

The first point is that it is always best to stay ahead of the game. If there are three days in which to ship an order it should not be left until the third day unless some unforeseen problem has occurred.

Many businesses offer a two-tier service with next-day delivery available at a higher price. This includes well-known examples such as major mail order companies, but also the car trade, where dealers typically receive a daily vehicle off road (VOR) delivery, perhaps guaranteed before 8am, and a weekly stock order. They pay higher prices for parts delivered under the former scheme.

I would make the following general points when planning:

- For fixed time deliveries, such as slot time deliveries for a major retailer, liaise with the person making bookings to avoid clashes.
- Plan for major promotions, ensuring labour and MHE is available.
- Consider using a wave picking system, under which picks are released to the warehouse in distinct batches. Many WMS have functionality to support this, and it is particularly helpful if coordinated with a vehicle loading schedule. An example is shown in Table 5.3.

Table 5.3 A fictional example of a wave picking schedule

Wave number	Description	Time issued	Completion time	Vehicle departure time
1	Non urgent orders	08.00	10.00	N/A
2	Non urgent orders	10.00	12.00	N/A
3	Non urgent orders	12.00	14.00	N/A
4	Uncompleted non urgents from previous days	14.00	15.00	N/A
5	Next day Northern Ireland	15.00	15.30	16.30
6	Next day Scotland	15.00	16.00	17.00
7	Next day North East	16.00	17.00	18.00
8	Next day North West/North Wales	17.00	18.00	19.00
9	Next day South West	18.00	19.00	20.00
10	Next day South Wales	19.00	20.00	21.00
11	Next day South East	20.00	21.00	22.00
12	Next day East Anglia	21.00	22.00	23.00
13	Next day local 1	22.00	23.00	07.00
14	Next day local 2	23.00	24.00	08.00

- Ensure that there is planned picking of less urgent orders, so this can be done before they become urgent. It may be possible to schedule this in the mornings before urgent orders are received, for example.
- Ensure that a non-urgent order becomes recognized as urgent if not shipped early. For example, if it must be shipped within three days it should be given urgent status on the third day.

- Do not prioritize to an extent that part of the operation is completely neglected: 'We never seem to get around to dealing with returns' is sometimes heard, but is not good management.

As with so many aspects of warehouse management, a well-planned picking sequence will aid a much smoother operation.

Picking receptacles

In some operations, eg in ecommerce, goods will be pre-packaged in the boxes in which they will be dispatched, with only re-labelling required. These can be picked in this form and placed in a roll cage, trolley with mesh sides, pallet with collar, or similar receptacle as the picker moves around the warehouse. Alternatively, in a small operation, a picker may in practice place picked items loose onto a trolley, or even carry them, and take them to a packing area in this form.

However, in most larger warehouses this will not be realistic, and the picker will need to place the items onto or into a suitable receptacle. This will certainly be the case if using an order picker, and is especially important if cluster picking to prevent elements of different orders getting mixed up with each other.

If the outbound shipment is to be palletized, it may be most efficient to place the cartons directly onto the pallet, to be wrapped when the order is complete. Placing a collar on the pallet may be a good idea to stop cartons falling off before it is wrapped.

Using the box in which goods will be dispatched is superficially attractive, and a WMS may be able to determine the size of box required for a particular order. However, in practice – especially if items are of different shapes and sizes – they are unlikely to be picked in the right order to be able to stack them into the smallest possible box. The result is that either the box will overflow, with risk of damage, until the pick is complete and it can be restacked properly, or oversized boxes need to be specified and will need packing out with a space filler. At the risk of stating the obvious, space fillers cost money, and big boxes cost more than small boxes.

It will often be best to pick into a temporary receptacle, a plastic tote box being the most common solution. Goods can be placed in the tote box whilst picking, then transferred to the box for dispatch, the tote being reused for the next pick. Whilst this represents an extra step in the process, it will often be the most efficient method in practice.

Dispatch

As with all other aspects of warehouse management, the key to efficient dispatch lies in the planning. Picking should be completed in good time, as should the proper preparation of the goods. This may involve packing, or even in some cases assembly if the customer requires this.

When these tasks have been completed the goods should be taken to a dispatch area laid out for the purpose. In a very small stores operation, this may simply mean that a couple of empty roll cages are provided, into which parcels are loaded during the day ready to be wheeled out to the parcels van when it arrives to collect. At the other extreme, a large operation shipping perhaps 100 full loads of palletized goods each day should have areas marked out of the correct size for 26 pallets each (assuming 1,200 x 1,000 millimetre pallets), this being the maximum number which can be fitted onto an artic.

> If mixed loads are shipped, perhaps with a mix of pallets and parcels to be trunked overnight on numerous vehicles to different parts of the country, I would recommend separation by physical barriers rather than relying on floor markings. The exact layout will depend on the nature of the operation.

If paperwork is generated elsewhere, which may be the case with export documentation, for example, there needs to be a robust communication system to ensure it is available on time, and to ensure that it is given to the correct vehicle driver. Without the correct documentation the consignment will most likely be detained in customs and create many difficulties in obtaining its release.

Finally, drivers should in my opinion be given the opportunity to check their load, and required to sign documentation accordingly. Some companies do not allow drivers on the loading dock for health and safety reasons, but if the driver has not seen the goods being loaded it can be difficult to enforce a claim if loss or damage occurs in transit. Most WMS will produce shipping documentation, which should be signed by the driver, also printing his name. A note should be made of his vehicle number and the date and time of collection. If goods are of high value they could be asked to provide photo ID, and in some very high-security environments the haulier is required to provide biometric data relating to authorized drivers before they are permitted to make a collection.

It is rare, but not unknown, for unauthorized drivers to make collections with the objective of theft. I have known a handful of such attempts during my career, most of them thwarted either when someone became suspicious, or when the culprits found the load unsaleable and it was found abandoned. However, in one case an unauthorized driver, almost certainly with the help of inside information from someone, managed to steal a full artic load of washing powder – an easily saleable and untraceable commodity. The crime was never solved.

Conclusion

As mentioned before, the warehouse process does not start with receipt, and by the same token it does not end with dispatch. There are other issues to consider, such as proofs of delivery.

However, dispatch seems a convenient point to end this particular chapter, and I will delve into such matters later in the book under transport operations.

Warehouse management systems

06

I have mentioned that I worked in a warehouse during my gap year in 1979, before the widespread availability of computers, and about half my time was spent working as one of five stock control clerks. We had a stock control card for each of our tens of thousands of part numbers. When a (handwritten) order was passed over from sales, we found the right card, and wrote by hand date, customer, order number, our advice note number, and number of units ordered; and deducted this by mental arithmetic from the remaining stock on the previous line, to give the new remaining stock. The reality was that although everyone tried their best, and at the time results were felt to be acceptable, by modern standards it did not work very well – an error of a thousand units could be made by the stroke of a pen or simple arithmetical error.

Today, much more help is available in the form of warehouse management systems (WMS), which can make major contributions to reducing errors and improving productivity. I would not advise anyone today to attempt to run a warehouse based on manual records, nor by use of spreadsheets: this is not the purpose for which they were designed.

Costs can vary enormously. A top-of-the-range system, with major bespoke development and customization to meet the specific needs of your particular organization, can cost millions of pounds, dollars or euros.

> At the other extreme, free WMS software can be found on the internet, such as ABC Inventory Software (Almyta, 2018). I have never used the system myself and therefore cannot testify as to its qualities, but from the online description it seems to provide the basic functionality required to manage a small warehouse. It does not appear to be suitable for a network in its free format, and there is no support package – although this is available if you buy a single user licence for $435.

Between the two extremes there are many options. Software Advice (2018) lists and reviews 94 different WMS. In addition, many enterprise resource planning (ERP) systems have a WMS module that can be purchased as an add-on so if, for example, your business is currently using Sage to support its accounting functions, then Sage Dexterity would be an obvious candidate to assess for suitability as a WMS.

If you are part of a large (perhaps multinational) organization, there may already be an agreement in place to provide a WMS module at a favourable price. It is important to ensure that the WMS is suitable for use in your area. For example, some systems developed in the United States do not have functionality for Value Added Tax (VAT), which will create problems in European Union countries. It would probably be better to source a suitable package locally, even if the cost is greater than that available for a comparable package sourced through the Global Procurement Agreement.

The main factors to consider are:

1. Can the WMS be interfaced with other system(s) currently in use within the business, or if necessary with systems used elsewhere, for example by customers or suppliers?
2. What functionality do you require and/or would like to have, and how much of this will the WMS provide?
3. Will the WMS support the type of automated data collection you wish to use – barcode, voice picking, etc?
4. How well will it work in practice?
5. What will it cost?

The final decision will almost certainly be a trade-off, deciding whether there is a sound business case to support additional expenditure in exchange for a better system.

To assist in making that decision I will now delve further into the various elements.

Interfaces

Almost every business is now system-dependent, and the days when multiple entry of the same data, or even paper trails, were thought acceptable are long gone for all except the smallest enterprises. For example, purchase orders should be visible to the warehouse (though possibly with sensitive price information hidden) so that they can verify that goods which arrive match that order; and the accounts department will wish to confirm that that those goods were received in good condition and that they should therefore pay the invoice.

If using the warehouse module of a company-wide ERP system such as Microsoft Dynamics, such internal interfaces should be standard (although it is worth checking, especially if the new module is a newer release than the other parts of the system). If you are using an in-house or legacy system for other activities but making an investment in a WMS, interfaces may need to be written: this will inevitably increase both the cost and the lead-time of the WMS project.

Externally, your customers may insist on sending orders electronically. Whilst some companies such as Travis Perkins in the UK still offer the option of issuing purchase orders by fax (Travis Perkins, 2018), others such as Volvo do not and the use of electronic data interchange (EDI) to the Volvo standard is mandatory (Volvo, 2017). In either event, a seamless electronic communication is by far the best alternative.

Similarly, you may wish your suppliers to communicate by EDI. To receive an advanced shipping notice (ASN) by fax or as a PDF attachment to an email is possible, but will require a manual data entry. This will incur a labour cost and at some point data entry errors will inevitably occur. It is far better to receive the ASN in an agreed format directly into the WMS.

There is also the question of communication with carriers. If you are shipping a few consignments per day a manual system will still work effectively. However, if you are working in an ecommerce organization shipping hundreds or even thousands of parcels each day to customers' homes, it is not feasible to expect a parcels carrier to manually input thousands of addresses and other details into their own tracking system. Some WMS systems such as Descartes are configured to interface with major parcels carriers such as DPD, Hermes and DHL (Descartes, 2018), which is a significant advantage.

Finally, I have made an implied assumption in this section that you will be running your own warehouse. If a 3PL is involved this will be a further complication, as they may well have their own preferred WMS package. Interfaces between this and your own systems packages should be an important factor in selecting a suitable 3PL.

Basic functionality

Any WMS should have basic functionality, including:

- receiving;
- put-away;

- pick;
- dispatch;
- inventory control;
- stock checking;
- statistical reporting (How many of X are currently in stock? How many of Y have been shipped to customer Z in the last year?).

Without these functions a WMS will not be able to operate effectively in most situations and I would recommend that it be rejected.

Advanced functionality

All except entry level WMS will be able to offer more than the basics, and some are very impressive. As a general rule, more expensive systems will have more features (Figure 6.1), and the key to the selection process is in deciding which functions are worth the additional expenditure in your particular organization. For example, if you need to track your products individually by serial number, this will be an essential feature. If you never need to do so, it will be an irrelevance. Other functions may

Figure 6.1 Launchpad screen for the numerous warehouse processing functions of an advanced WMS

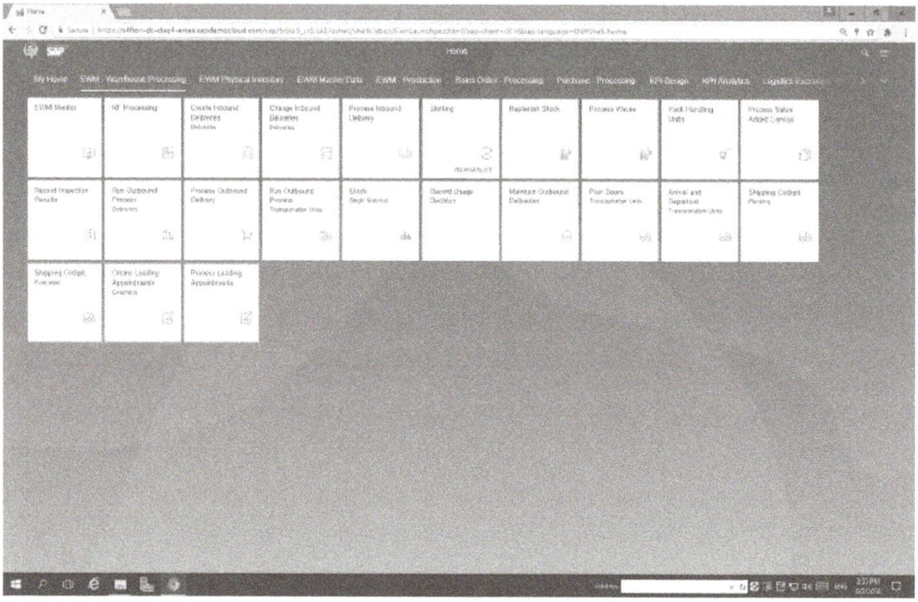

SOURCE Reproduced by kind permission of SAP (UK) Ltd

come into the 'nice to have' category, and a judgement call may be required as to whether the cost is justified. The list of possible functions is almost endless; a selection of them are described below.

Yard management and delivery slot time allocation

In a medium size organization this may well fall into the 'nice to have' category. An advanced system may not only be used to manage bookings of delivery slots, but will also allocate inbound loads to the best dock (Figure 6.2). For example, if there is an area of cantilever racking for out-of-gauge items at one end of the warehouse, the WMS will route such material to a dock closest to that area. It will also recognize drop trailers as a special category. Clearly this sort of functionality is much more useful on a site with 60 docks than with two docks.

Pre-allocation of stock

In a simple system, stock must be received and put away before it can be allocated to an order. However, in a slightly more advanced system it is possible to pre-allocate, or set up the WMS to do so automatically (Figure 6.3).

Figure 6.2 Slot time allocation function. The screen shows details not only of the individual booking but also an overview of capacity available and in use throughout the day

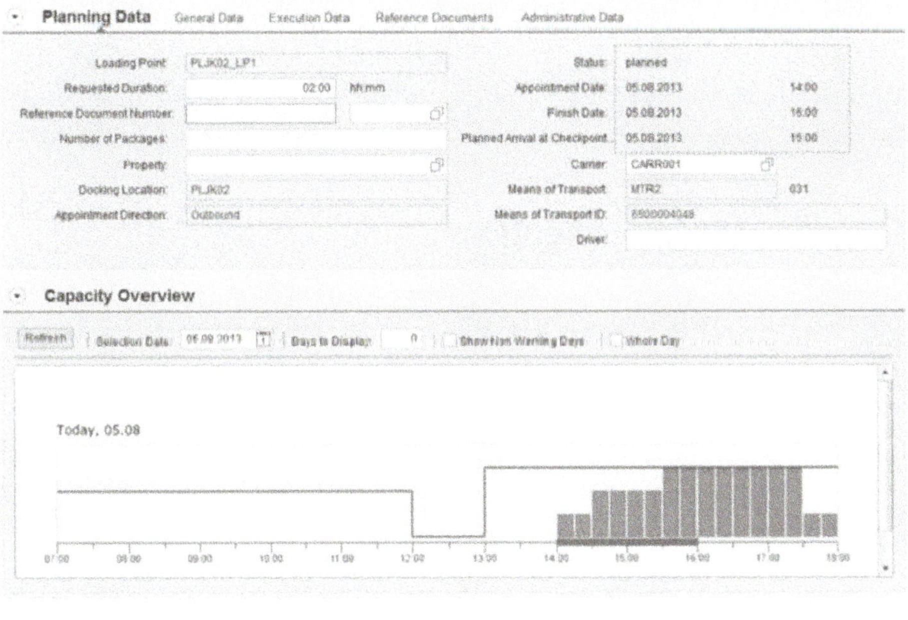

SOURCE Reproduced by kind permission of SAP (UK) Ltd

Figure 6.3 Stock summary, showing pre-allocation of 44 units of the fifth item to an as yet unpicked outbound order

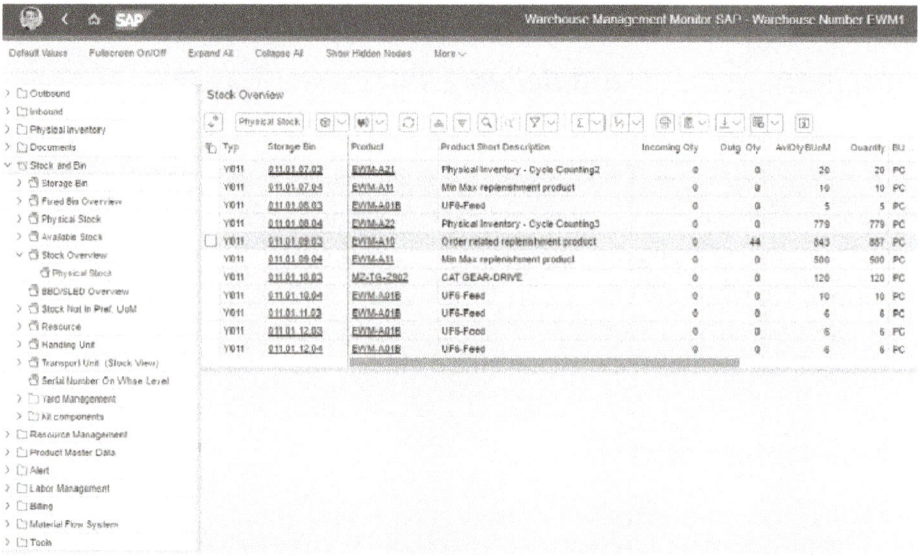

SOURCE Reproduced by kind permission of SAP (UK) Ltd

In advanced systems, parameters can be set – should part of an incoming consignment be retained for future emergency orders rather than being used to fulfil non-urgent back orders? A very advanced feature is known as opportunistic cross-docking. This will entail inbound stock being allocated to an outbound order the same day to save putting away and retrieving that stock, even though the same items may already be in storage. Some may not wish to use such a feature anyway as it is not compatible with first-in, first-out (FIFO) but it illustrates the sophistication of some packages now on the market.

Allocation of storage locations

Most systems will be able to generate a location into which an inbound item should be put away (Figure 6.4). In very basic systems this may be random, but usually a WMS can be pre-programmed with parameters such as an ABC analysis, or certain products to certain areas. However, keeping reserve stock above the picking face for that particular item may be beyond the capabilities of some products on the market.

Figure 6.4 Availability of empty locations for put-away

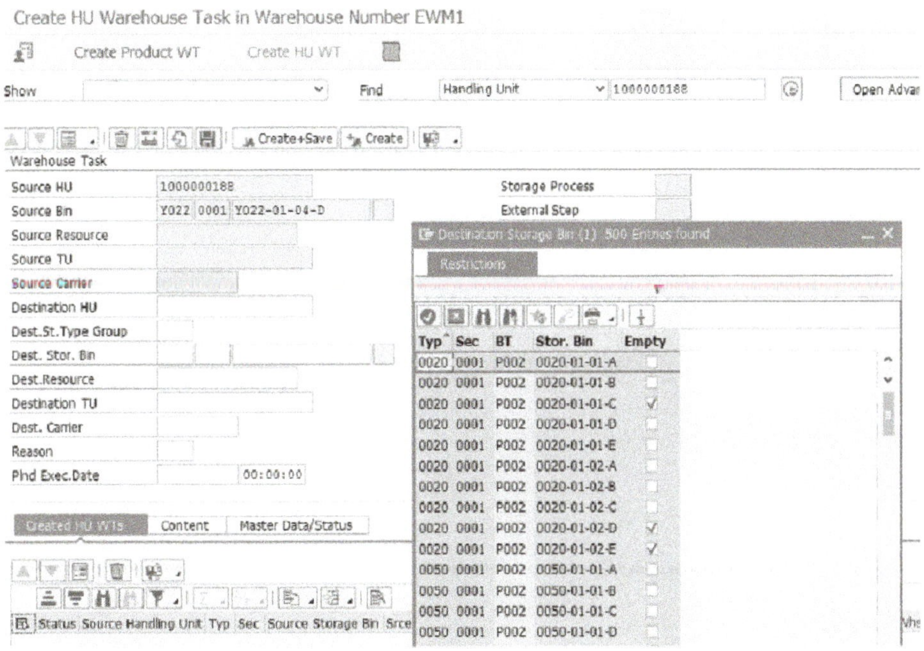

Create HU Warehouse Task in Warehouse Number EWM1

Create Product WT | Create HU WT

Show | Find: Handling Unit | 1000000188 | Open Advar

Create+Save | Create

Warehouse Task

Source HU	1000000188
Source Bin	Y022 0001 Y022-01-04-D
Source Resource	
Source TU	
Source Carrier	
Destination HU	
Dest.St.Type Group	
Dest. Stor. Bin	
Dest.Resource	
Destination TU	
Dest. Carrier	
Reason	
Plnd Exec.Date	00:00:00

Storage Process

External Step

Destination Storage Bin (1) 500 Entries found

Restrictions

Typ	Sec	BT	Stor. Bin	Empty
0020	0001	P002	0020-01-01-A	
0020	0001	P002	0020-01-01-B	
0020	0001	P002	0020-01-01-C	✓
0020	0001	P002	0020-01-01-D	
0020	0001	P002	0020-01-01-E	
0020	0001	P002	0020-01-02-A	
0020	0001	P002	0020-01-02-B	
0020	0001	P002	0020-01-02-C	
0020	0001	P002	0020-01-02-D	✓
0020	0001	P002	0020-01-02-E	✓
0050	0001	P002	0050-01-01-A	
0050	0001	P002	0050-01-01-B	
0050	0001	P002	0050-01-01-C	
0050	0001	P002	0050-01-01-D	

Created HU WTs | Content | Master Data/Status

Status | Source Handling Unit | Typ | Sec | Source Storage Bin | Srce

SOURCE Reproduced by kind permission of SAP (UK) Ltd

Promotional packs, kitting and bills of materials

Each of these is, in logistical terms, a very similar thing. A wine merchant may sell mixed cases with a particular selection of wines; a medical supplies company may sell items individually or in particular quantities and combinations in a range of first aid kits; or a solar energy installation may require a bill of materials consisting of panels, fittings, and electrical components. In each of these examples pre-selection of the items, and assembly into the pack or kit, is necessary before shipment (Figure 6.5). Not all WMS provide the necessary capability; and work-arounds involving 'shipping' the components then 'receiving' the kit are far from ideal.

Packing, sub-assembly and added value activities

A similar principle applies when changes need to be made to the product to meet customer requirements. These range from labelling, repacking or re-palletization; through inserting batteries; to the uploading of complex software – and there may be more than one step in the process. Again work-arounds are possible, perhaps shipping an item as XYZ, labelling and reboxing, then receiving as XYZ QVC. If such activities are occasional this may be acceptable, but shortcomings such as losing track of the product during the process will inevitably lead to problems at some point.

Figure 6.5 Screenshot showing a bill of materials, in this case all components needed for the production of 348 Model M525 bicycles

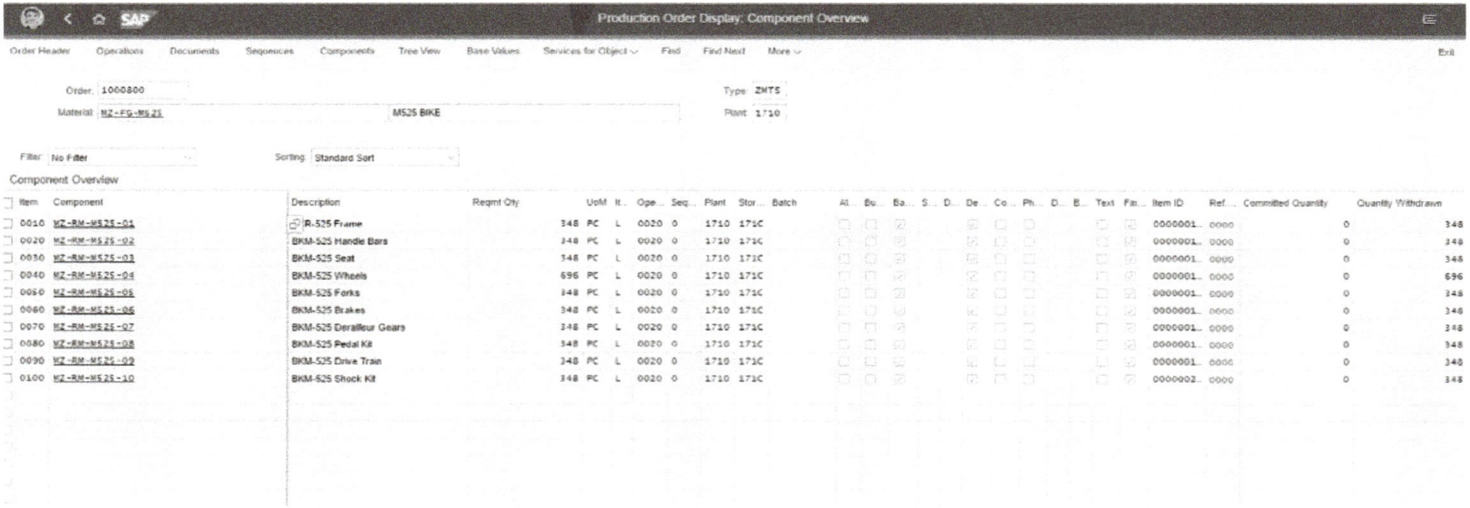

Production Order Display: Component Overview

Order Header | Operations | Documents | Sequences | Components | Tree View | Base Values | Services for Object | Find | Find Next | More | Exit

Order: 1000800 | Type: ZMTS
Material: MZ-FG-M525 | M525 BIKE | Plant: 1710

Filter: No Filter | Sorting: Standard Sort

Component Overview

Item	Component	Description	Reqmt Qty	UoM	It...	Ope...	Seq...	Plant	Stor...	Batch	Al...	Bu...	Ba...	S...	D...	De...	Co...	Ph...	D...	B...	Text	Fin...	Item ID	Ref...	Committed Quantity	Quantity Withdrawn
0010	MZ-RM-M525-01	R-525 Frame	348	PC	L	0020	0	1710	171C														0000001...	0000	0	348
0020	MZ-RM-M525-02	BKM-525 Handle Bars	348	PC	L	0020	0	1710	171C														0000001...	0000	0	348
0030	MZ-RM-M525-03	BKM-525 Seat	348	PC	L	0020	0	1710	171C														0000001...	0000	0	348
0040	MZ-RM-M525-04	BKM-525 Wheels	696	PC	L	0020	0	1710	171C														0000001...	0000	0	696
0050	MZ-RM-M525-05	BKM-525 Forks	348	PC	L	0020	0	1710	171C														0000001...	0000	0	348
0060	MZ-RM-M525-06	BKM-525 Brakes	348	PC	L	0020	0	1710	171C														0000001...	0000	0	348
0070	MZ-RM-M525-07	BKM-525 Derailleur Gears	348	PC	L	0020	0	1710	171C														0000001...	0000	0	348
0080	MZ-RM-M525-08	BKM-525 Pedal Kit	348	PC	L	0020	0	1710	171C														0000001...	0000	0	348
0090	MZ-RM-M525-09	BKM-525 Drive Train	348	PC	L	0020	0	1710	171C														0000001...	0000	0	348
0100	MZ-RM-M525-10	BKM-525 Shock Kit	348	PC	L	0020	0	1710	171C														0000002...	0000	0	348

SOURCE Reproduced by kind permission of SAP (UK) Ltd

Figure 6.6 Warehouse operative's handset showing menu for picking options. This handset has a rugged design to protect it against damage in warehouse use

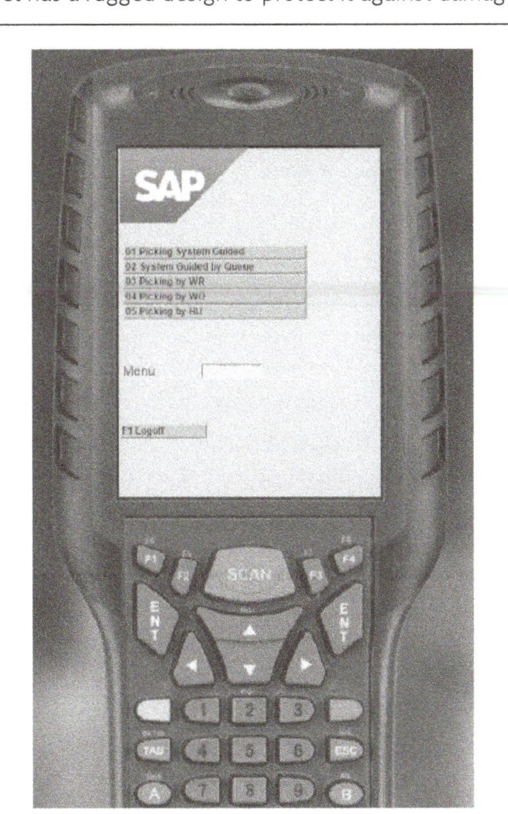

SOURCE Reproduced by kind permission of SAP (UK) Ltd

Picking

I have described earlier various strategies for picking, such as batch and wave picking. With simpler systems, picks are generated individually and it will be the job of the supervisor to allocate them to individual pickers. A more advanced system will generate waves or batches of picks if required, and sequence them in accordance with parameters such as vehicle dispatch times (Figure 6.6).

It may even have a task interleaving function, by which for example a forklift driver would be tasked with putting away a pallet, then retrieving a pallet from a nearby location, so that the forklift is loaded in both directions to and from the shipping and receiving area. This is undeniably a good idea, and to operate such a system perfectly would be a Utopian ideal, but it rarely works quite as well in practice as might be hoped.

Automated replenishment

In some businesses there is a policy of automatically replacing sold stock (subject to manual intervention where necessary), and if this is the policy within your organization the WMS will need to support this. At the very least it should be capable of producing detailed statistical reports of past usage, to provide the necessary data to predict future usage, and thereby enable the placement of purchase orders in the correct quantities.

Customs warehousing

If you wish to operate any form of bonded warehouse in which goods are held free from payment of duty and/or VAT, you will not be granted permission to do so unless your WMS has specific functionality to support this.

Labour and resources

This is another area in which the level of functionality can vary enormously, from zero upwards. Mid-range systems will aid the warehouse manager in deploying resources for different tasks, for example allocating operatives and forklifts in line with vehicle arrival schedules. The information will also be useful in determining the overall resources required – on a busy day overtime may need to be worked and/or agency staff recruited (Figure 6.7). Performance monitoring, both of the operation as a whole and of individual operatives, is a useful management tool.

It will also be useful to compare personal performance, for example of pick rates. Often a firm word, and the knowledge that they are being monitored, will be sufficient to increase the productivity of an under-performing worker.

The most sophisticated packages will be interfaced with the human resources and payroll functions (Figure 6.8). Holidays and overtime can be controlled by the warehouse manager and entered directly onto the system, which will then make the necessary payments directly to the worker and avoid any need for a separate system of time monitoring.

Dispatch and transport

Outbound orders need to be managed in the same way as inbound orders, for example allocating orders to vehicles and to dispatch bays (Figure 6.9).

Many WMS have no functionality beyond dispatch, but there is an opportunity for further added value in transport planning. In a simple case this may involve

Figure 6.7 Overall daily performance of a warehouse showing inbound, outbound and overdue pick statistics. Clearly this was a quiet day for inbound movements

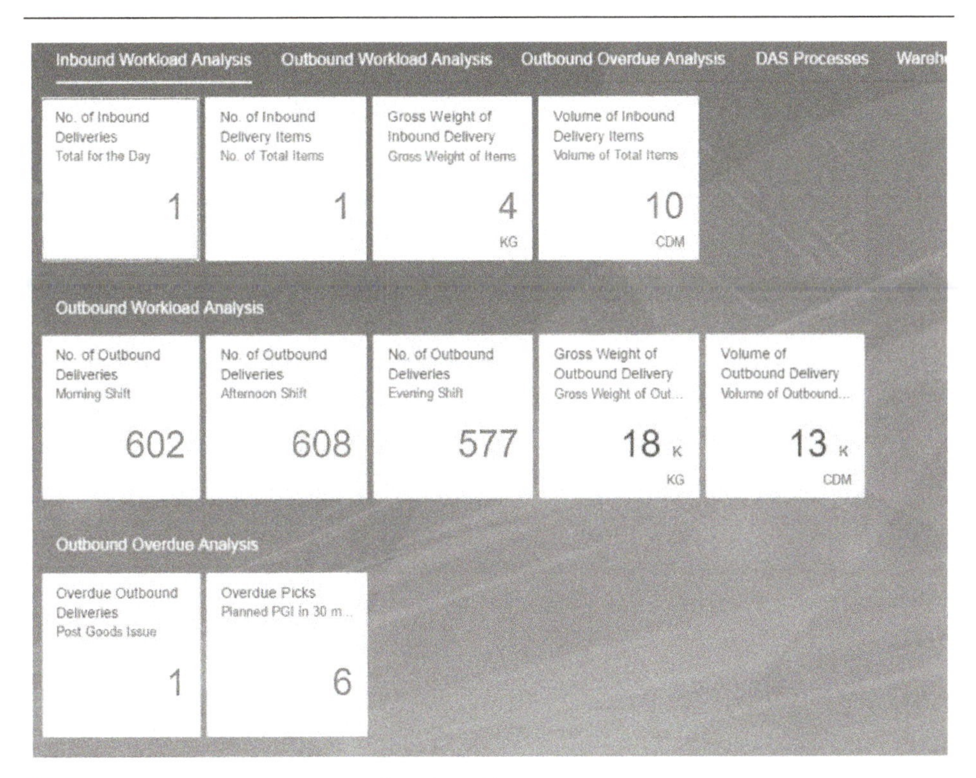

SOURCE Reproduced by kind permission of SAP (UK) Ltd

Figure 6.8 Time and attendance functions of a warehouse labour management module

Menu
Show Hidden Nodes

Warehouse Management Monitor SAP - Warehouse Number LM01

- Outbound
- Inbound
- Physical Inventory
- Documents
- Stock and Bin
- Resource Management
- Product Master Data
- Alert
- Labor Management
  - Labor Utilization
  - Executed Workload
  - Indirect Labor Task
  - Planned Workload
  - Shifts
  - Time and Attendance
- Billing
- Material Flow System
- Tools

Time and Attendance

Processor	Description	Event Date	Time	Event...	Event...	Team Le...	Name Lead	Proc.Group	Name Grou...	External Data
BROWN	William Brown	14.04.2017	07:54:00	I	WORK	GARCIA	Carlos Garcia	LM01TEAM1	LM01 Team1	
			12:02:00	O	BREAK					
			12:59:00	I						
			17:01:00	O	WORK					
GARCIA	Carlos Garcia		07:59:00	I						
			12:01:00	O	BREAK					
			12:58:00	I						
			[illegible]	O	WORK					
GOMEZ	Esmeralda Gom...		07:56:00	I						
			12:02:00	O	BREAK					
			12:59:00	I						
			17:03:00	O	WORK					

SOURCE Reproduced by kind permission of SAP (UK) Ltd

Figure 6.9 Outbound order management, showing amongst other data size of orders, door allocation, and real time picking status

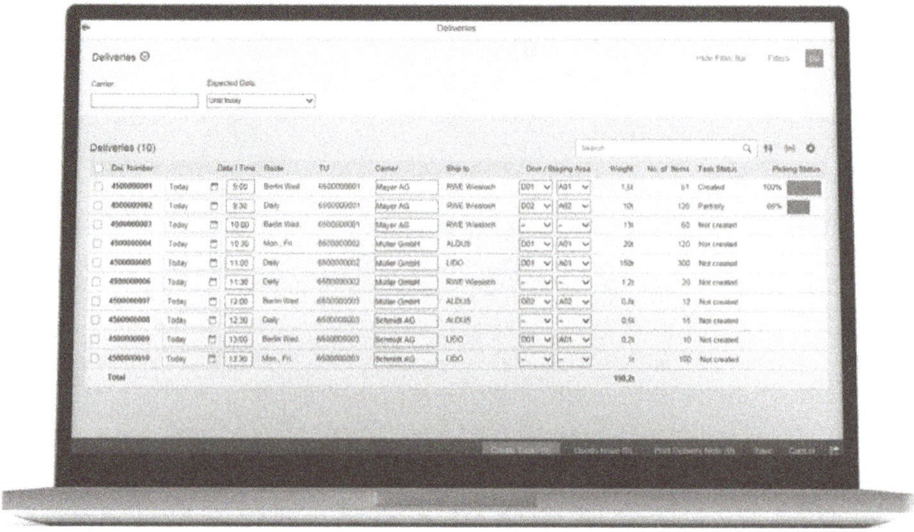

SOURCE Reproduced by kind permission of SAP (UK) Ltd

purely geographical allocation – parcels for England and Wales may be delivered by carrier A, those for Scotland by carrier B, etc. A more advanced system will be able to select the best or cheapest carrier for a particular consignment according to geography, consignment size and delivery time requirements – one pallet for next day delivery to Paris will be sent via carrier C, on a three-day delivery via carrier D, or six pallets on a three-day delivery by carrier E.

The most sophisticated systems will include a route planning facility, so that for example a consignment of six pallets for Cologne and one of nine pallets for Frankfurt can be loaded on the same vehicle, which can yield substantial savings in transport cost.

Finally, a good system should include a function to properly manage proofs of delivery, whether on your own or 3PL vehicles.

Automated data collection

We are all familiar with automated data collection (ADC) in our daily lives – a barcode scanned at the supermarket checkout to identify the item we are buying as a bottle of diet cola, or at the entrance to the theatre to verify that our print-at-home ticket has not already been used by someone else. In the former case we see the immediate effect of the data capture as the cola is added to our bill at the right price (based on a centrally managed database), and the sale will be part of the process to trigger a replenishment delivery from the distribution centre to the store, and ultimately a delivery from the supplier. The information will probably also be used as part of the monitoring of the performance of the checkout operative, and when we scan our loyalty card, to provide marketing information.

A similar principle will apply in the warehouse environment. The barcode will identify the particular item, but the data captured from it will be used by the WMS in myriad ways.

It is possible to operate a warehouse without any form of ADC – in practice this will mean paper-based processes within the operational area and manual entry of information to the WMS in the office. For a very small warehouse, perhaps with a single operative, this will be adequate, and issues such as the absence of real time information may be unimportant. Indeed, it might be difficult to justify expenditure on anything more technically advanced – though life can be made easier by simple things such as providing a clipboard.

However, for a medium-sized warehouse there are clear advantages in using ADC, both in accuracy and productivity, and for a major operation it is essential.

A WMS must be selected that is able to support the particular method of ADC chosen, and the two decisions (selecting a WMS and selecting an ADC methodology) are therefore interconnected. In this section I will describe the main options available.

Barcode

Many products will already have a printed barcode, and in some circumstances this can be used directly. If you have ordered a printed copy of this book from a mail order company there is a good chance that the barcode on the back cover will have been scanned by the picker to ensure that they have picked the right book.

However, in most cases one or more separate barcodes will be used within the warehouse. A barcode may be applied to the pallet on arrival; there may be barcodes on the racking to denote locations; and a barcode may be added on dispatch to track the item. I will discuss barcode printing in the next chapter.

In many warehouses a handheld scanner will be used. The scanners usually seen at the supermarket checkout are not designed for industrial use, and I would strongly recommend what is known as a rugged scanner. (Even though some mobile phones have a scanning function I would certainly recommend that this be seen only as an emergency backup solution). Basic models function only as a scanner, but more advanced models have a keyboard and display screen so that the operator can access the WMS directly (Figure 6.10). This is usually preferable, as a manual override may be necessary, as in some cases is a partial manual entry (for example a forklift driver may be able to scan a barcode at ground level to indicate a pallet location but needs to input '5' to indicate that he has placed the pallet on level 5).

Older scanners tended to work on a stand-alone basis, and would require downloading via a modem. Modern versions operate via Wi-Fi connections and are thus

Figure 6.10 Powerscan by Datalogic SRL barcode scanner being used for automated data capture

SOURCE Reproduced by kind permission of Datalogic SRL

able to update the WMS in real time: as a pallet is scanned into a location it will immediately increase the available stock.

Scanners can also be desk-mounted, either portable or fixed. The former is useful if scanning items near dispatch, as both hands are available to hold the product; and the latter can be placed alongside a conveyor to scan each item as it passes without human intervention, perhaps as part of an automated sortation system.

The biggest growth area is in wearable scanners. The main body of the device is usually worn on the wrist, with a small scanner worn on the finger. They are light in weight and therefore take just a short time to get used to, and save the time normally expended on picking up and putting down a scanning gun – Richards and Grinsted (2016) estimate that a maximum pick rate of 150 items per hour can be achieved with a wearable scanner, compared with a maximum of 100 per hour with a gun.

> The main advantage of barcode scanning is in accuracy – far less concentration is required than with a paper system. The disadvantages are that a barcode must be applied to every picking unit, and that there must be a clear line of sight between the scanner and the barcode. Applying labels and maybe turning items around to scan the barcode will take time, and in terms of direct time spent actually picking, a barcode system may be no faster than a paper-based one.

However, accuracy will be much improved, a substantial benefit in itself, and time lost due to resolving discrepancies will be avoided. I would therefore recommend barcode ADC as superior to a paper system in most applications.

RFID

RFID stands for Radio Frequency IDentification, and involves data capture by radio waves. It is used in applications such as security access cards and toll passes, but the same technology also has applications in logistics.

The commonest form is passive RFID, each tag being able to be read only by a suitable RFID scanner. They hold limited data but can be interfaced with a more comprehensive database, as with a barcode. They are in use by some retailers such as Marks and Spencer, and if shopping in such a store you may have seen RFID tags attached to garments. A survey has revealed that, since their introduction, retailers have improved stock accuracy by 50 per cent, reduced total stock holding, and actually increased sales by between 1.5 and 5.5 per cent due to improved product availability (Internet Retailing, 2018).

The main advantage over barcodes is in speed. To scan the barcodes on every hanging garment on a rail would require handling each individually; with RFID one can simply run the reader along the length of the rail. Richards and Grinsted (2016) estimate that pick rates of 200 to 300 items per hour are obtainable using RFID. Similarly, Keller and Keller (2014) estimate that RFID gives a 58–63 per cent improvement in productivity over barcode systems.

The main disadvantage is the cost of the tag – around 10–20 pence/cents (2018 prices). The cost of a barcode can be minimal, especially if it is incorporated into printed packaging. This can be a substantial difference over a year's sales. There can also be problems with missed items – when running the reader along the rail, one can not be certain every tag was read; and the read range is short: RFID tags at height cannot be read from ground level. Also, metals and liquids interfere with signals, so it is not suitable for such products.

An active tag is also available, which has its own power source, holds a much greater amount of information, and can be updated. These can cost up to £40/$50, which would preclude their use in all but very high value items. They can, however, be used on totes or roll cages that remain within the logistics chain, and used to track items en masse as, for example, they pass a fixed reader at a transhipment centre, the tracking system automatically updating for all items in that roll cage.

Figure 6.11 Pick by light operation

SOURCE Reproduced by kind permission of SSI-Schaefer. © Copyright SSI-Schaefer

Pick by light and put to light

These are related methodologies, the common factor being that an operative is drawn to the appropriate location by a light.

There are obviously variations of both options, but it may be easiest to describe a typical pick by light scenario (Figure 6.11). Small parts are held loose in plastic tote boxes on a picking face, replenished by gravity feed from behind when empty. On the shelf beneath each tote box position is a light, a numerical display, and up and down buttons. The picker will scan a barcode indicating which order they are about to pick, and lights will come on simultaneously to indicate all of the locations they will need to visit for that order. When they reach the first location, the number will indicate the quantity of that item to be picked. If the quantity is, say, six and there are six or more units in the tote, they will pick them, place them into a picking receptacle, and press the down button six times. If there are only four in the tote, they will pick them, press the button four times, remove the empty tote to allow the next to move forward, then pick the remaining two and press the down button twice more. When the display reaches zero the light will go out, and when all lights have been extinguished the pick is complete. A variation is to have just a single button to press when the pick is made, which is quicker but does not force the picker to think about quantities – and this is especially difficult if they have forgotten how many they still need to pick from a replenishment tote.

Sometimes a confirmatory step is included in the process, scanning a barcode on the item as it is picked. This will improve accuracy but reduce speed.

The main advantage of pick by light is indeed speed. Richards and Grinsted (2016) estimate 250–450 items per hour to be achievable. This is faster than any other non-automated and non-semi-automated method, and in the right situation it can work very well in practice. Disadvantages include the length of time taken for implementation, and the cost: Richards (2018) estimates €150,000 for a 1,000 location installation. It is not suitable for batch or cluster picks, or for multi SKU locations, and contingency action in case of a breakdown can be very difficult.

Put to light, or pick to light, is more suited to cluster picking. Typically, a pick is made, and the items taken to a packing area. In that area are an array of boxes or totes, with a light and numerical display. The operative scans a barcode to indicate that they are ready with item X. Lights will illuminate to show all the boxes requiring item X, the operative will place them in the box in the right quantity, and press the relevant button to extinguish the light. When no lights remain visible (which should coincide with a zero quantity remaining) the operative can move on to the next SKU. This solution will not be appropriate for all applications, but in a suitable situation it can work very well.

Voice

This technology had its origins in the 1990s, and was first used in cold stores where writing and paper systems were especially difficult to use due to the thick gloves worn by operatives (Figure 6.12).

In essence, the methodology is very simple. An operative will be equipped with a headset consisting of earphones and a microphone. They will receive an instruction such as 'Go to location 123', and on arrival they will say 'At location 123'. The next instruction will be 'Pick six pieces of item XYZ'; they will do so and say, 'Picked six pieces of item XYZ'.

Most versions have a multilingual function. The warehouse at which I first saw this technology happened to have a large Polish contingent amongst the workforce, and they were able to both receive instructions and respond in their native language.

The main advantage of voice technology is simplicity – it is easier and quicker to train operatives than with other methods. It is completely hands-free, so both hands can be used for handling goods. Productivity is moderate: Richards and Grinsted (2016) estimate 100–250 items per hour, which is better than paper or barcode but slower than RFID or pick to light. Accuracy is also good – Keller and Keller (2104) estimate that errors can be reduced by at least 80 per cent compared to paper-based methodologies. Disadvantages include the need for a quiet environment, and difficulty in accommodating mixed SKU locations or serial number tracking.

Figure 6.12 Voice picking in cold store. The thick gloves are necessary due to the extremely low temperature, but they make paper-based picking very difficult

SOURCE Reproduced by kind permission of SSI-Schaefer. © Copyright SSI-Schaefer

Augmented reality or vision picking

Whilst related to the concept of head up display (HUD), used in applications such as military aircraft for several decades, this is a new technology in the warehouse environment.

Operatives are provided with special glasses. Within the field of vision is displayed an instruction of a similar nature to that used in voice technology. The glasses also incorporate a barcode scanner so that the location and/or the item can be scanned merely by moving the head, and a visual or audible alarm is triggered if the wrong barcode is scanned. Completion of the pick is confirmed by tapping the side of the glasses or by pressing a special device worn like a wristwatch.

Ubimax (2017) say they installed a pilot for such a system with five sets of equipment at the Samsung Parts Centre in Breda, Netherlands, in 2015. This was then increased to 30 sets in 2016. They report a 22 per cent increase in the speed of picking for fast-moving items, and a reduction of 10 per cent in errors.

It is too early to make definitive statements about the long-term future of vision picking, but it certainly shows promise.

Other technical possibilities

Visual recognition is an established technology, used for example in automated car number plate recognition. Its usage in logistics is at present mainly limited to the reading of addresses in mail sorting applications. It would technically be possible to use this technology to eliminate barcodes, and read part numbers or even features such as book covers directly. However, development costs would be substantial and perhaps the benefits would not justify expenditure.

Finally, I understand that navigational information is now available for operatives within warehouses – for example, move three aisles forward, turn left, and access the fourth pallet space on the right. My personal opinion is that this will only be of benefit if the warehouse layout is confusing and/or training of operatives is poor. My recommendation would be to resolve these problems rather than investing in this particular technology.

Choosing a system that will work well in practice

> No matter how good your WMS, it will not turn around bad warehouse performance. If you have underlying issues such as poor housekeeping, pilferage due to lax security, overcrowding due to obsolete stock or simply bad attitudes, you need to solve these problems first.

Most new WMS installations work well in practice, and people usually notice a rapid and dramatic improvement. Having said that, there are sometimes teething problems, and occasionally there are disappointments when a system falls short of expectations. Careful preparation will greatly improve the chances of getting the best out of a new WMS.

I would suggest the following:

1. Find out what all stakeholders would like to achieve from installing a new WMS – those directly affected such as operatives; internal stakeholders such as finance and the sales team; and if appropriate customers and suppliers.
2. Use this to draw up a list of outcomes you want to achieve. You may have ideas about how you want to achieve those outcomes, but be prepared to listen to alternatives, as there may be a better way.
3. Look for a system that is intuitive, and which avoids counter-intuitive entries. (For example, one system offers a drop-down box including 'return', but if you want to process a return, you select the other option!).
4. Get buy-in from senior management at an early stage, and manage expectations in terms of cost, making sure there is a clear business case for investment. To be told at the end of the process to come up with an alternative solution at half the price is not what you want to happen.

5. Make sure the potential supplier has experts in warehousing who will understand your needs, and not just IT experts and a good sales team.
6. Try to avoid too much bespoke development work. This will cost money and add to project lead-time, and there is a risk that there will be genuine misunderstandings and differences in interpretation over the specification. Also, there may be teething problems with the newly developed parts of the system, and you may not want to be the guinea pig.
7. Try to source the whole package from a single source. I was once in a situation where the system did not work, the hardware supplier blamed the software, the software supplier blamed the hardware, and the customer was getting impatient. Do not let this happen to you.
8. Check online reviews, take up references with existing users of the system, and preferably visit them and talk to hands-on users who may be more forthcoming about problems. Look for users whose warehousing needs are similar to yours – the particular industry is less relevant than WMS functionality.
9. Ensure that help and support will be available whenever it might be needed – 24/7 if that is the nature of your business.
10. Plan well in advance so you don't need to rush implementation.
11. Make sure you will get on with key people on a personal basis. It will make a big difference during the implementation phase.

Cost

Finally, I come to the question of cost. As with so many other purchases, costs are not restricted to a one-off cost of purchasing the software. The elements to take into account will include the following:

- Software. Obviously, there will be a headline cost of the WMS software itself but will upgrades be necessary elsewhere in the company's ERP system? Will there be ongoing licence fees, and will they include software updates? Does the licence cover a limited number of users?
- Customization. If possible, agree a fixed cost for this in advance. Charges on a per day basis can quickly escalate.
- Hardware. This should include not just terminals for desk-based users but mobile equipment too. Will terminals be provided on forklifts or order pickers? Will barcode or RFID scanners be needed, including spares in case of breakdown or increased demand? Will the Wi-Fi coverage in the warehouse need to be improved?

- Support. Is there an additional charge for calling the provider's helpdesk? Will the company need to train a super-user and how much of their time will need to be dedicated to this task?
- Implementation. Does the cost of the software include on-site support from the supplier's personnel during implementation? Will time and input from the company's own IT department be needed?
- Training. When will staff be trained on the new system? Will replacement labour need to be brought in to cover? Will external trainers need to be used, and if so at what cost?
- Temporary downturns in productivity. With the best will in the world, operatives will take time to get up to speed with any new system. Any additional staff who need to be brought in during the launch period should be taken into account.
- Future proofing. It is worth thinking ahead to predict any future changes. Perhaps the business is expected to grow so more user licences will be required, and it is possible that some functionality not required at the outset will be needed in the medium term. It will be easier to negotiate a good price for an option on such items at the outset – when you have the upper hand in negotiations – rather than when the system is installed, and the supplier knows that they have the upper hand in negotiations.

Conclusion

As mentioned at the beginning of this chapter, a WMS of some sort is almost essential in a modern warehouse. The cost may appear large, but improvements in accuracy and productivity will usually more than justify this cost. In many cases, users soon come to think that the improvement since installing a new WMS is so great that they wonder how they ever got by without it.

Packaging and customer requirements

07

So far, I have discussed how to set up and run a warehouse, up to the point of dispatch. However, when an item is dispatched you will need to be sure that it will reach the customer in the condition in which they would wish to receive it. If they are a large organization, they may make some very specific demands – I will expand on this towards the end of this chapter. However, all customers will expect that the goods have been properly packaged so that they arrive in good order, and adequate packaging forms part of the terms and conditions of most carriers: the old excuse of 'It was all right when it left here' will cut little ice these days.

I will therefore begin with some general comments about packaging.

Packaging: General principles

The idea of packaging dates back to at least 8,000 years BCE, when clay pots were used for storage and/or transport of grain, olives and oils (Emblem and Emblem, 2012).

Packaging is something we come across every day, but few give it a second thought, and indeed most people probably do not even stop to think why we package goods.

The answer is that packaging has four main functions:

- To contain the product (ie keep all the elements together). This may mean that a clutch kit for a car will comprise the cover, plate and bearing in the same box; or that 1 kilogramme of frozen peas is contained within a plastic bag.
- To sell the product. This is most important in the retail sector, where packaging is designed to entice the consumer to buy the product, and this will rarely be a logistics task. However, some retailers insist that products are presented in a particular manner, which could be a logistics task.

- To provide information. In the retail sector, the primary packaging (see below) will probably carry details such as nutritional information, which is unlikely to come under the remit of logistics; but outer packaging will need to carry labelling such as descriptions, quantities, barcodes and delivery address, which may well do so. For some industrial products (eg a metal bar or pallet of roof tiles), the packaging will be the only source of information.
- To protect the product. This is usually the function with which logistics operations are most concerned. Most products will need to be boxed, crated, palletized or otherwise packaged to protect them from:
  - impact damage (including puncture);
  - crushing;
  - vibration;
  - deterioration due to light, heat, cold, humidity, vermin, microbial attack;
  - tampering or theft.

This is perhaps also the appropriate point to introduce three definitions used in the retail sector, as the terms will be used in the text:

- Primary packaging. This is the packaging taken home by the consumer. For example, in the case of a packet of pain-killing tablets, it will include the blister packs that are in direct contact with the tablets, and the box containing the blister packs.
- Secondary packaging. This is the packaging used to group products for ease of handling – for example, 24 cans of vegetables may be contained within a cardboard tray that is placed on the supermarket shelf.
- Tertiary packaging. This is packaging used for ease of transport, such as a pallet with stretch wrap, or a roll cage. This aspect of packaging will be that with which a logistics operation is most likely to be concerned, and more often than not goods for delivery to a retailer will need to picked and then placed in tertiary packaging prior to dispatch.

All three classes of packaging can be seen in Figure 7.1 – the bottles and jars visible in the roll cages are primary packaging; the boxes secondary packaging; and the roll cages themselves tertiary packaging.

Similar principles apply to mail order retailing. It is most likely that goods will have been stored in their primary packaging, but will need to be placed into a larger box before dispatch, which will fulfil the functions of both secondary and tertiary packaging.

Opinions vary as to the point in the supply chain at which packaging should take place. Pålsson (2018) is an advocate of packaging postponement, leaving such operations until as late a stage as possible. This has the advantage that the risk of unnecessary or duplicated effort is minimized. Transport costs and risks of in-transit

Figure 7.1 The different classes of packaging used in the retail sector

SOURCE Reproduced by kind permission of Crown Lift Trucks Ltd (www.crown.com)

damage can also be reduced – for example if wine is shipped from Australia in bulk in a tank container but only bottled on arrival in Europe. However, the cost of both labour and materials will, for example, be cheaper in Far Eastern countries than in Western Europe, and this difference may favour packing at an earlier stage. As so often in logistics, the best advice is to consider both options and decide each case on its individual merits.

EU packaging waste regulations

There has always been a trade-off in terms of the level of packaging used. Both cost and environmental considerations dictate a reduction in the amount of packaging used, but if this results in poorer protection, there may be more wastage due to product deterioration or damage, which is also undesirable from both the cost and environmental perspectives. The aim must be to use packaging in a smarter manner rather than simply increasing or decreasing the amount used.

This is now enshrined in UK law. The Packaging (Essential Requirements) Regulations 2015 dictate that the amount of packaging used must be the minimum necessary to ensure the safety, hygiene and acceptability of the product. It should also be recoverable by recycling, energy recovery, composting or biodegradation.

There are specific regulations regarding the manufacture of reusable plastic pallets, and glass packaging. A full copy of the regulations can be found online (UK Government, 2015).

The Producer Responsibility Obligations (Packaging Waste) Regulations 2007 apply to organizations with an annual turnover of more than £2 million, and who handle more than 50 tonnes of packaging per year. They must make annual data submissions regarding the type and quantity of packaging used, and produce evidence that they have recovered a certain amount of packaging waste, and recycled a percentage thereof. Schemes under which the obligations of several producers are grouped together may be established subject to certain conditions, and in practice obligations are sometimes met by financing the recovery and recycling of other businesses' waste. A full copy of these regulations can also be found online (UK Government, 2007).

These UK laws were enacted due to EU directives and equivalent laws will therefore apply throughout the EU. Other countries have their own regulations or actively regulated voluntary codes, including Australia and many US states. It is specifically stated that the EU law applies to all packaging – primary, secondary and tertiary. There are therefore few large businesses outside of the purely service sector to which the laws do not apply.

The regulations appear to be successful, as in the UK more than 80 per cent of packaging is recycled, and 76 per cent of packaging is made from recycled materials (Emblem and Emblem, 2012). Unless you are a small company and therefore exempt, you should familiarize yourself with, and of course follow, these regulations.

Primary packaging

Primary retail packaging comes in a huge variety of formats – cans, drums, glass bottles, plastic bottles, plastic trays, plastic pots, and boxes of all shapes and sizes. Normally the primary responsibility will rest with marketing and product development activities – ensuring that the product is appealing to the customer and suitable for use in the home. Whilst a logistics operation may be tasked with re-boxing or a similar activity, a product will usually be received into a warehouse already in its primary packaging. This is therefore outside the scope of this book.

However, there may be an impact on logistics, and this should be taken into account. For example, cooking oil and washing powder are often sold in bottles or boxes of square or rectangular cross-section. This improves space utilization in transit by over 25 per cent compared to a cylindrical product, but whisky and washing-up liquid are rarely sold in this format. There are undoubtedly further opportunities to both reduce costs and help the environment that could be exploited.

Also, secondary and tertiary packaging may derive part of their strength from the product and its primary packaging. Examples include butter, where the hydrostatic pressure within each primary pack gives additional strength to the secondary pack;

and wine, in which the bottles give strength to support the case above. To use a thinner and inadequate glass in wine bottles would be a false economy.

Corrugated board

> Commonly known as corrugated cardboard, this is the mainstay of the packaging industry. In the UK alone 2.1 million tonnes of paper is used each year to manufacture 4.1 billion cubic metres of board (Emblem and Emblem, 2012).

'Cardboard boxes' are indeed the mainstay of the packaging industry, and the most obvious solution for expendable packaging. That is therefore where I will start.

There are four types of corrugated board (Figure 7.2):

- Single faced. This consists of a single corrugated and a single flat layer. It is flexible but does not have great strength. It is therefore used mainly for direct wrapping of products.
- Single wall. This has one corrugated and two flat layers. It is the most widely used general-purpose board for both boxes and trays.
- Double wall: This has two corrugated and three flat layers and is therefore stronger than single wall. Uses include boxing of large appliances and furniture, long-term storage and display stands.
- Triple wall. With three corrugated and four flat layers, it is extremely strong and sometimes used as a substitute for wooden crates. It is used for heavy-duty applications such as metal automotive components, eg gearbox covers.

Figure 7.2 The four main types of corrugated board: single face, single wall, double wall, triple wall

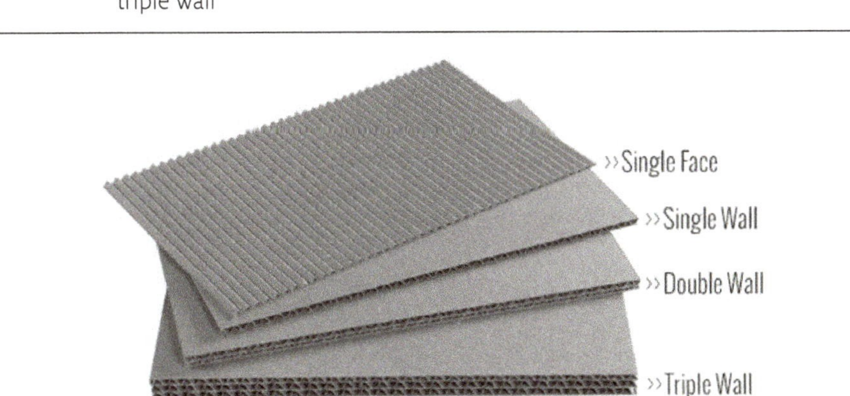

SOURCE Reproduced by kind permission of Macfarlane Packaging

New boxes are invariably supplied in flat-pack form – delivery costs would otherwise be prohibitive. If you are running a small business you may be able to meet some of your requirements by recycling inbound packaging, but you should make sure that it still has enough strength and rigidity. Few small business customers will complain – and if they do you can justify this policy on environmental grounds. In either case the box should be securely sealed with good quality tape: a minimum 48mm width is recommended by some carriers. Skimping by using just a single length of tape, or using domestic Sellotape, is a false economy.

> One item no warehouse can do without is a tape gun. If you are opening a new warehouse do not forget to buy at least one with a good supply of tape. (If you are unfamiliar with tape guns, there is one resting on the box in Figure 7.18.)

If goods are to be protected by tertiary packaging, or shipped on your own vehicles and handled carefully, standard boxes will be sufficient. However, specialist postal packaging is recommended for mail order, or shipment via any parcels or shared user network. Postal boxes are designed with tuck-in flaps to eliminate lines of weakness that could be opened up by rough handling and vibration in transit (Figure 7.3).

Smaller items such as books or games can be shipped in a corrugated board postal wrap, which will usually have a self-sealing strip.

Figure 7.3 Specialized postal box, designed to eliminate lines of weakness

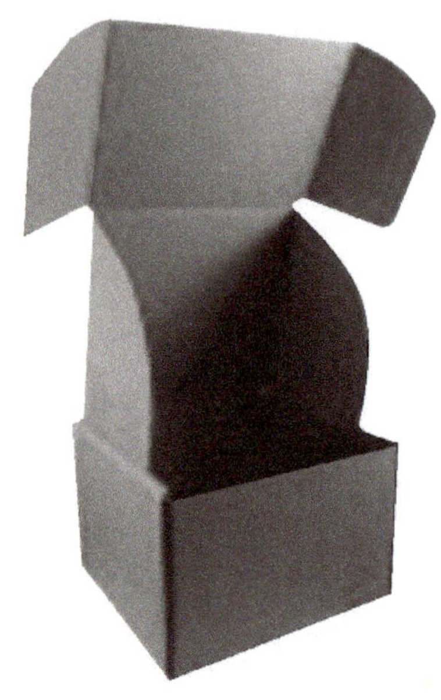

SOURCE Reproduced by kind permission of Macfarlane Packaging

Most box manufacturers will market a standard range of box sizes and shapes. However, some suppliers will offer a bespoke service and produce boxes to meet your specific requirements (for example, if you sell a large number of a particular product that does not fit a standard box).

You can also purchase machinery to manufacture your own boxes on site, making a range of boxes and inserts from flat corrugated board. This will obviously require substantial investment, and will not be viable unless you are producing at least 200 boxes per day. It has the advantage that boxes can be produced to the exact dimensions you wish, and printed with your own or customer's logos and instructions such as 'Handle with care'. Printing on the inside of boxes is also possible, with personalized messages such as 'Happy 37th birthday to my darling wife Yvette'.

Case erector machinery will take a flat pack box, fold it into shape and seal it by tape or glue (Figure 7.4). Manufacturers can supply semi-automatic or fully automatic versions, entry-level equipment requiring a four-figure investment: different models will obviously have different capabilities with respect to minimum and maximum box sizes. An alternative is machinery that will simply tape an already filled and erected box. Whilst this expenditure will only be justified if there is a sufficient throughput of boxes, making up cartons manually and sealing with a tape gun is one of the most tedious and unpopular jobs in the warehouse, and you will probably make your operatives very happy if they are spared this task.

Figure 7.4 Semi-automatic carton erector machine

SOURCE Reproduced by kind permission of Robopac

Figure 7.5 Postal cube. This is supplied in a flatpack roll but is square in cross-section, and therefore makes better use of space in transit and is not prone to rolling around

SOURCE Reproduced by kind permission of Macfarlane Packaging

It is common to ship some items, such as wall posters or long protective strips, wrapped in a cardboard tube. Unfortunately, cylindrical tubes do not make best use of transport space, and can have an annoying tendency to roll around, especially on conveyors. Alternatives worth considering are 'postal cube' which is supplied as lay flat tubing, but is square in cross section when opened out (Figure 7.5), or triangular tubes, offered for sale for example by DHL to their customers.

If the product will potentially be damaged by moisture, such as electrical goods or the inside of a briefcase, silica gel packs will absorb water vapour from the air and prevent a build-up of humidity.

Voidfill

Common sense dictates that items should not be allowed to rattle around loose in a box, and therefore some form of filler will be required to prevent this. The technical name for this material is voidfill. It has the additional benefit of impact and crush protection, and a fragile item such as a vase should always be placed in an oversized box with voidfill. Voidfill should always fill the box – empty space will increase the risk of crush damage.

It may also be a good idea to wrap items, for example in bubble wrap or mesh wrapping; to use dividers to stop items knocking against each other (eg wine bottles in a case); and to place loose items in a plastic bag.

The materials in common use as voidfill include:

- Sealed air pillows. As the name implies, these are air filled plastic pillows. They can be purchased in a ready inflated form, but if being used in quantity it is worth considering purchasing the pillows as a roll, together with a machine to inflate the pillows as they are needed (Figure 7.6).
- Bubble wrap. This works on a similar principle, and is available with different bubble sizes to suit different sizes of product.
- Fanfold paper. The paper is purchased flat, but pre-folded so that it can be opened out into a fan shape which makes an effective cushion.
- Shredded paper.
- Crumpled paper.
- Packing chips. These are usually about 2 centimetres long and used as loose voidfill. Options include expanded polystyrene and biodegradable chips. It should be noted that some of the latter are made from products such as cereals and are edible by rodents. If using these it is important to keep them in a sealed container and clean up spillages to avoid attracting vermin.
- Bespoke polystyrene packing. Televisions, computers, etc, are usually packed in purpose-designed polystyrene mouldings that fit exactly to the product's dimensions, and to the internal dimensions of the box. These are mass-market items where the development cost of the bespoke packaging can be spread over very many units, but in very high-value markets such as defence or medical diagnostic equipment that development cost can be justified even for a much smaller number of units.

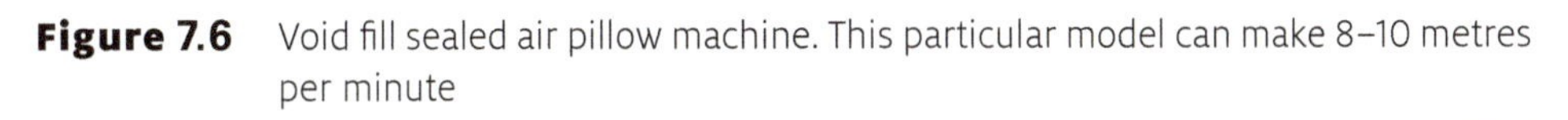

Figure 7.6 Void fill sealed air pillow machine. This particular model can make 8–10 metres per minute

SOURCE Reproduced by kind permission of Rajapack

In all cases, re-use of clean voidfill from inbound consignments when packing for dispatch is highly desirable for both cost and environmental reasons.

Other expendable packaging

Whilst corrugated board is the most common form of expendable packaging, there are alternatives.

The simplest alternative is the padded envelope or bubble mailer, ie an envelope that incorporates a layer of bubble wrap or other protective filler. These are available from high street retailers and will be familiar to most people.

Also at the smaller end of the spectrum, plastic bags are an option. These are often used for items such as magazines, folded garments, and printer cartridges. Standard sizes such as A4 and A5 can be bought with self-seal flaps for just 3 or 4 pence/cents, which is likely to be cheaper than alternatives. Transparent plastic is satisfactory for brochures or low-value items, but I would recommend opaque plastic as a theft deterrent in most cases.

Larger or irregular sizes can be wrapped using rolls of lay-flat plastic tubing, which is cut and sealed as appropriate using a heat sealer. This is available in varying gauges or thicknesses, 62.5 microns being recommended for medium duty use, such as a soft toy; 125 microns for heavy duty use, such as a rug or roll of carpet; and 250 microns for extra heavy duty – this is often used for export applications where repeated rough handling can be foreseen (Kite Packaging, 2018). A basic manual sealer can be purchased for a few tens of pounds/dollars/euros, but I would recommend at least an industrial hand-operated sealer priced at a few hundred, which will make a wider seal and stand up to more arduous use. It will typically have a capacity of perhaps three to four bags per minute. For higher volumes a continuous automatic unit should be considered: the bag is fed into one end and drawn through the sealing unit (Figure 7.7). At the top end of the market fully automated systems can handle hundreds of bags per hour.

For larger and heavier items, palletized cartons are an option (Figure 7.8). These usually consist of a corrugated board sleeve and lid, pre-attached to a wooden pallet. They can be used to house a number of smaller boxes, or a single irregular item: at least one famous manufacturer imports motorcycles in this format. The pallet base permits forklift handling.

A variation on this idea is the pallet collar (Figure 7.9). These consist of wooden boards of the same length as the sides of a standard pallet, usually hinged at the corners to allow them to fold flat when not in use. They can be placed around the edge of the pallet, so as to form a collar, typically about 20 centimetres high. Several collars can be stacked on top of each other if necessary, and a lid used on top of the uppermost collar. Smaller cartons or irregular shaped goods can be placed inside, and the collars give sufficient strength and rigidity for the pallet to be stacked as well as directly protecting the contents. My one word of warning is to ensure you have a system in place to monitor their return – if they regularly go missing and need replacement this will soon become expensive.

Figure 7.7 Automatic continuous heat-sealing machine

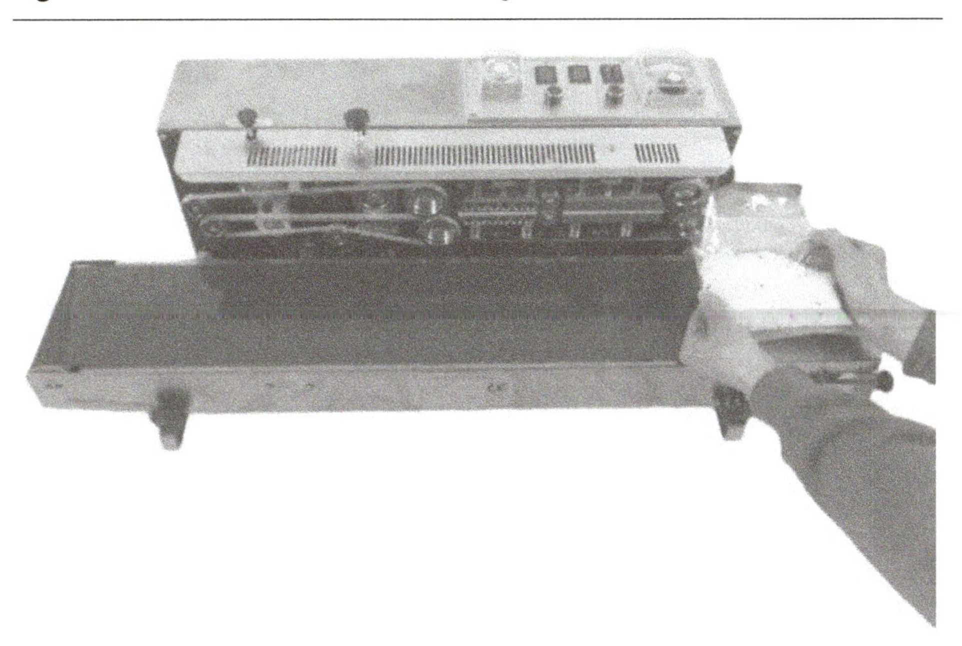

SOURCE Reproduced by kind permission of Yolli Ltd in the UK

Figure 7.8 Palletized carton

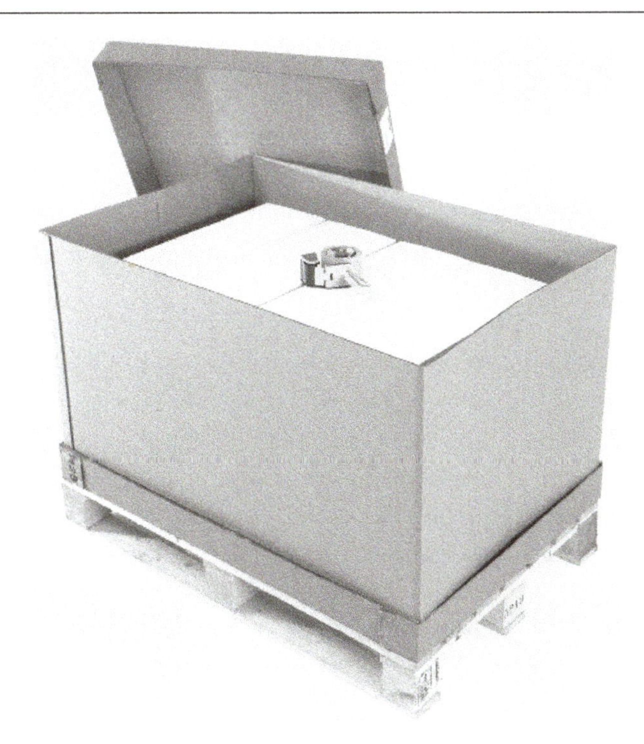

SOURCE Reproduced by kind permission of Davpack

Figure 7.9 Pallet collars

SOURCE Reproduced by kind permission of Davpack

The final option I will mention is the wooden crate. These can be purchased in foldable plywood form suitable for medium duty applications. However, the heaviest of items, such as major aircraft or ship components, should be packed in specially constructed hardwood crates, with a sufficient thickness of wood to prevent any risk of crushing or collapse.

Whether a crate or palletized carton is used, goods must be secured inside it. For regular shaped items such as boxes or reels, packing the empty space with wood or other suitable material may be sufficient – such material is known as dunnage. For irregular and/or inherently unstable items – such as a motorbike or coil of steel – constructing a specially designed cradle and/or bracing with straps to the floor of the crate will also certainly be necessary (Figure 7.10).

All timber packaging should be treated by heat or chemicals to ensure that it is free from insects and fungal infection. Some countries have strict biosecurity controls in place that apply to timber imported as packaging just as much as they do to timber imported in its own right. For example, for goods arriving by airfreight into Australia, timber packaging or dunnage not meeting the Department of Agriculture and Water Resources offshore treatment requirements must undergo methyl bromide fumigation, kiln drying, heat treatment, ethylene oxide fumigation, Gamma irradiation, or a full unpack and inspection at an approved arrangement site. Each of these will attract a fee (Australian Government, 2018). It is wise to familiarize yourself with regulations for both testing and documentation before making a shipment to such countries.

Figure 7.10 A bespoke wooden shipping crate

SOURCE Reproduced by kind permission of Barnes & Woodhouse

Pallets

Most people in the logistics industry will have seen countless pallets over their careers – they are ubiquitous.

If you are delivering your own product on your own vehicles, there is no reason why you should not choose a pallet size to best suit your own needs. Indeed, if you have a product such as an engine or generator that does not readily fit a standard size pallet, that may well be the logical course of action.

Having said this, standard sizes have evolved for good reasons. The commonest size in use in the UK is 1,200 x 1,000 millimetres; in Europe it is the Euro-pallet, 1,200 x 800 millimetres. These both have the advantage that they fit well onto the 2,420 millimetre wide standard artic trailer: other sizes will give poorer space utilization. Common sizes in use elsewhere include 1,100 x 1,100millimetre, which is widely used in Asia; 1,140 x 1,140 millimetre, which fit well into the standard 2,330 millimetre internal width of a seafreight container; and 1,219 x 1,016 millimetre, based on Imperial measurements, which is the most common in the USA.

It is also easier and cheaper to buy standard sized pallets, and they are available through pallet rental schemes. They will also be more acceptable to customers.

You may at this point be thinking that your carton sizes do not fit neatly onto standard pallets. If that is the case, I am sorry to say that my advice would be to change your carton size if possible. To have an overhang will risk damage, and will create difficulties stowing pallets economically on a vehicle. In fact, some carriers will go as far as to refuse to carry pallets with overhanging cartons. If changing carton size is not feasible due to the dimensions of the product, I would recommend stacking them inside a palletized carton or pallet collars (see earlier section).

> Something else to avoid is the pyramid stack. This often has its origins in companies using cheap and weak cartons. They find that pallets arrive damaged, and blame the carrier for stacking items on top. To prevent this, they stack cartons in a pyramid so that nothing can be stacked on top. The result is that the carrier is obliged to ship fresh air above the pallet. Many carriers impose a surcharge in these circumstances (DHL, 2018) and they are, in my opinion, absolutely justified in doing so.

Most pallets are still made from wood – this is the traditional material, and it has stood the test of time.

Plastic pallets are also available, which can be thoroughly steam cleaned and therefore are preferred by some, for example in the food industry, for hygiene reasons. However, they can be slippery. This means that goods must be secured more tightly to the pallet, and there have been accidents involving a sudden stop by a forklift resulting in the smooth plastic pallet sliding off the end of the smooth metal forks. Greater care is therefore needed when using plastic pallets. Metal pallets are less common, but are used by some tyre companies, for whom a wooden splinter in their product could have serious consequences.

Another alternative is to use pallets made from recycled wood chips bound with resin. These often have feet that are moulded into an almost pointed shape, which tend to damage the pallet below when stacked. I personally do not like them for this reason. A final option is cardboard pallets: these are suitable only for very light duty applications.

It is also worth thinking about the construction of pallets, as this varies. First, pallets can either be two-way entry, meaning that forks can be inserted only from front or back; or four-way entry, meaning that forks can be inserted from any side. The former may be stronger and more durable, and many Euro-pallets are built this way, with solid wooden runners along either side and along the centre line – a stack of two-way entry pallets can be seen to the right of the shipping crate in Figure 7.10. However, there have been cases where a vehicle has been loaded from the side with two-way entry pallets, with the result that on arrival the customer has been unable to offload from the rear. This is an embarrassing mistake to make.

Also, a pallet may be constructed with bottom perimeter boards, ie the base has boards around all four sides. This again aids strength and durability, and if they will always be lifted by forklift this would be my preferred option. However, such pallets are not suitable for use with some types of pallet stacker as one set of forks will lift and another remains at near ground level, with the result that the base perimeter boards will be stripped away as the pallet lifts. This will irreparably damage the pallet and pallet stackers should therefore only be used with pallets (such as Euro-pallets) without such boards. This does not apply to pallet trucks that lift the pallet only a few centimetres off the ground. It has been suggested that chamfered edges on base perimeter boards will make it easier to insert the wheels of a hand pallet truck. Whilst this is undeniably true, it is rare for anyone in a warehouse environment to make such a stipulation in practice.

Whichever type of pallet is used, it is of course essential to check that they are strong enough for the intended use, and in good condition. Any pallets found on inspection to be rotten should be discarded, and any damaged pallets either repaired or discarded.

Finally, pallets should be stored under cover, to prevent them getting wet and damaging goods; and securely. There is a market for used pallets, most of which of course is legitimate, but stolen pallets are difficult to trace and not difficult to sell.

Securing cartons to pallets

Most palletized consignments do not use collars or palletized cartons, so boxes will need to be secured to the pallet. There are two main methods of doing so: banding or stretch/shrink wrap.

Banding

Also known as strapping, this involves wrapping a number of steel or plastic bands around the pallet, each strap passing under the top of the pallet and over the cartons (it is also sometimes used to securely seal individual cartons) (Figure 7.11). Strap widths of 9, 12 and 16 millimetres are standard. If you are shipping a few pallets only, a hand-operated machine, which tensions the bands by pulling a lever, will be adequate. For higher volumes, a cordless hand-held electric tensioner will be preferred, and for very high volumes fully automated equipment is available. Whichever version is used, plastic or cardboard corner guards should be used to prevent the banding cutting into the top edges of the cartons.

Polypropylene banding is suitable for most products. For heavy-duty applications such as construction material or timber, polyester strapping can be tensioned more tightly. Steel banding is sometimes used, but edges can be sharp, and it is more difficult to remove on arrival. In all cases I recommend thick gloves and eye protection when using or removing strapping – I recall not doing so when I was a young man and learning a painful lesson cutting myself on steel banding.

Figure 7.11 Two palletized cartons with lids attached and secured by banding

SOURCE Reproduced by kind permission of Davpack

Plastic wrapping

This usually involves wrapping cartons onto a pallet using a plastic film, usually of polyethylene, which is at least superficially similar to domestic cling film. The most basic method is to use a roll of film, typically 400 millimetres wide, and a simple hand dispenser (Figure 7.12). The operative secures one end of the roll and circles the pallet repeatedly from the base upwards until all cartons have been enclosed.

Figure 7.12 A hand-held dispenser for stretch wrap – this particular model is a two-handled version

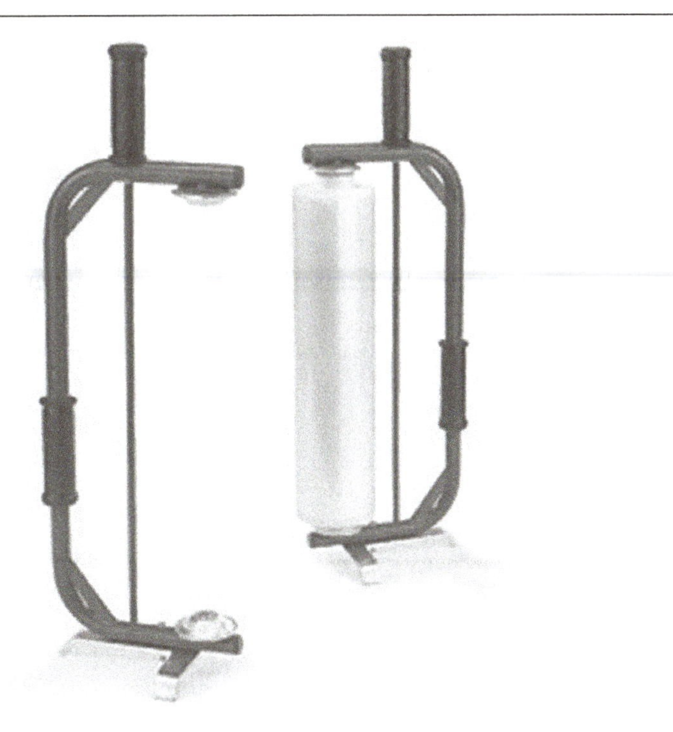

SOURCE Reproduced by kind permission of Rajapack

There are two common mistakes made by inexperienced operatives. The first is not to wrap securely enough. The film should be stretched tight, and each layer should overlap the layer below by about 50 per cent. A pallet may look adequately wrapped at first sight, but the vibration from any journey can cause the load to work loose, and may result in the collapse of the stack of cartons. The other is to fail to attach the stack to the pallet, wrapping only the cartons. The result is that stack of cartons is free to slide off the pallet en masse. It is important to ensure that the pallet base is included in the wrap – tying the free end of the roll to a corner post of the pallet is not really a recommended technique but in my experience it tends to work.

If you are wrapping large volumes of pallets, it can be tiring to hand-wrap and operatives can end up getting dizzy by continuously going around in circles. A semi-automatic machine may be a good investment in such cases. The basic type has a turntable, on which the pallet is placed (Figure 7.13). The free end of the roll is secured to the pallet, and at the push of a button the turntable rotates and the roll is dispensed, rising until a photoelectric sensor detects the top of the pallet, when it descends to wrap a second layer. Capacity can range from six to 40 pallets per hour according to the model.

Figure 7.13 Semi-automatic stretch wrapping machine of the turntable type

SOURCE Reproduced by kind permission of Robopac

If purchasing such a machine I would recommend also buying ramps, which are usually an optional extra. If pallets are always loaded onto the turntable by forklift these are not necessary but they do give the option of using a hand pallet truck. Installation by an experienced tradesman is also recommended.

Some models work slightly differently, with the pallet remaining static and the stretch wrap applied by a rotating arm, or by a robotic machine that circles the pallet. Fully automated machines usually include a conveyor to feed in pallets.

A variation is shrink-wrapping. In this case the plastic film will usually be of polyolefin, which shrinks when heated. This creates a good seal and is commonly used in the food industry, for example for meat and cheese. The film can be heated by a hand-held heat gun, or in automated systems by passing through a heated tunnel incorporated into a conveyor system.

> In most cases see-through plastic film will be used. However, black film may be a good idea for security purposes if the product is attractive to thieves, such as wines and spirits.

Pallet inverters

In some circumstances it may be necessary to change the pallet under a load. This can happen when a pallet is damaged, or when a different type of pallet is needed – plastic rather than wooden, or a pallet from a hire pool, for example. This can obviously be achieved by restacking each individual carton by hand, but that is highly labour intensive and not an activity one would want to carry out repeatedly.

The solution may be a pallet inverter (Figure 7.14). On the commonest models the loaded pallet is placed into the inverter, and the 'new' pallet is placed on top. The whole load is then inverted, ie turned upside down, and the 'old' pallet can then be removed from the top. The load on its new pallet is then ready to be removed from the inverter – re-wrapping or re-banding may be necessary. Some customers may object to receiving products 'upside down', even if in practice it makes no difference. (With some products, for example bags of cement, such objections are unlikely.) In these cases, re-inversion to the original orientation may be necessary. Some models work by tipping the pallet just beyond the horizontal, to enable removal of the 'old' pallet and its replacement in the same position, which avoids this problem.

Figure 7.14 A pallet inverter

SOURCE Reproduced by kind permission of Premier Pallet Systems Ltd

Other reasons for inverting pallets include removal of a damaged package from the bottom layer, and ensuring that corks are kept moist in bottles of fine wines.

Durable packaging

So far, I have talked mainly about expendable packaging (although pallets and wooden crates can of course be reused). In many situations there is no viable option: for an ecommerce business to arrange collection of large numbers of small reusable containers from individual private addresses will far exceed the cost of providing disposable cardboard boxes. (An exception is that of supermarket online deliveries, which are made on dedicated vehicles, and plastic containers are unloaded on arrival and returned on the same vehicle.)

Conversely, if you are shipping repeatedly to the same customers, or perhaps to your own network of retail outlets, then durable (ie reusable) packaging becomes a more realistic solution. Indeed, in some industries it is widespread. For example, public houses in the UK will receive draught beer in re-usable kegs, and bottled beers and mixer drinks in returnable glass bottles. However, those same products will be sold in supermarkets in expendable cans and bottles.

One general point to note is that opportunities to use durable packaging in both directions should not be missed. I know of one organization that shipped goods from the UK to Spain in durable containers, which were returned empty. Other goods were shipped from Spain to the UK in identical containers, which were returned empty! This organization is no longer in business and such practices cannot have helped.

Plastic tote boxes

The most common alternative to a cardboard carton is the plastic tote box. These come in a variety of shapes and sizes; with and without hinged lids; and with angled sides so as to be nestable and thereby save space when not in use, or with vertical sides (Figure 7.15). One well-known high street fashion retailer uses lidded tote boxes for deliveries of all except hanging garments, delivering to individual stores and returning empty totes at a later date to the central warehouse. As a dedicated vehicle fleet is used, there is no incremental transport cost for their return.

Euro-containers are popular, as the sizes of these are such that they nest upon each other, and fit in easy multiples on a standard 1,200 x 800 millimetre Euro-pallet; sizes include 300 x 400 millimetres and 600 x 400 millimetres, for example.

Dollies, ie wheeled trolleys, are available in sizes specifically designed for the movement of Euro-containers, and these are useful for movements within the warehouse.

Figure 7.15 A stackable plastic tote box. These are commonly used in a variety of sizes for storage and transport

SOURCE Reproduced by kind permission of SSI-Schaefer. © Copyright SSI-Schaefer

The main advantage of durable packaging is that there is no repeated expenditure on cardboard or similar materials; plastic boxes can last a very long time. Whilst figures vary widely, the one-off cost of a lidded tote box is likely to be around 20 times that of an expendable box of similar size, so if it is used once a week for a year this will represent a 250 per cent return on investment. They are also more robust so in-transit damage is likely to be reduced, and there are of course environmental benefits.

The main disadvantage is that a procedure must be set up for their return. In a closed system this will not be difficult, but if they are being delivered to a large number of customers they will need to be tracked to ensure the same number is returned as has been shipped. Not only is there a temptation 'not to be bothered', but plastic totes are an attractive item for theft: they will for example make an excellent container for children's toys in an under-stairs cupboard. There will also in some circumstances be a transport cost for their return, and in all cases a handling cost.

Metal stillages

These have been in use in the automotive, aerospace and manufacturing industries for decades (Figure 7.16). The major automotive manufacturers have a range of stillages, which they use internally for inter-plant movements, and may also provide them to suppliers for component shipments. Some individual stillages may be many years old, as they last a very long time. The greatest hazard to longevity is water – they should be kept dry, to avoid rust.

The main disadvantage of metal stillages is that they are heavy, and can only be moved by forklift. This will not be a problem in practice if they are used for heavy engineering components that would require mechanized handling in any event.

Most standard stillages are not collapsible, but if you are setting up a system for the first time, this option should be considered.

Figure 7.16 Mesh-sided metal stillages

SOURCE Reproduced by kind permission of Lowe RP Ltd

Roll cages

Just as stillages have been in use in the automotive sector for many years, so the roll cage has long been a retail standard. Examples can be seen in Figure 7.1, and most people will have seen them in grocery stores. Some have shelves and are intended to be displayed in the store: for example, my local supermarket uses roll cages of this type for milk. Most, however, are intended to be wheeled out into the store, contents unloaded and placed on shelves, and the roll cage returned to the store room and thence to the warehouse.

Their main advantage is that they can be wheeled around by hand without any form of mechanical assistance. They can be pushed on and off of vehicles via a dock leveller, or loaded and unloaded using a tail-lift or scissor lift. This is important when delivering to stores where forklifts are unlikely to be available. A disadvantage is that they cannot be stacked, so much of the free height of a trailer may be unutilized. One solution to this problem is to use a double-deck trailer – see Figure 8.21 (page 000) for an example.

Pallet and box hire pools

I have discussed expendable packing, whereby pallets and boxes are not intended to be returned; and durable packaging, which is intended to be returned to source for re-use. There is a third alternative, which is one-way hire. This can have both financial and environmental benefits.

The market leader in this field is CHEP (an acronym for Commonwealth Handling Equipment Pool), whose distinctive blue pallets are widely used in the retail sector. They have 220 locations across 35 European countries, and are also active in 19 other countries including the US (CHEP, 2018). Typically, a supplier to the UK supermarket industry will arrange a delivery of, say, 500 pallets into their premises (there is an option to collect). They will use these for shipments to supermarket distribution centres, maybe at the rate of 24 per load, recording the numbers and destinations on CHEP's system, and be charged a daily rental fee on each pallet according to the length of time it is held. When empty, the pallets will be collected from the distribution centre by CHEP, and returned to the system. They are of a very robust construction, and are designed to be re-used many times, after repair if necessary.

Gefbox operate a similar system, providing rigid plastic boxes to serve the automotive industry. Empty boxes are delivered to suppliers who use them for deliveries to manufacturing plants, from where they are collected, steam cleaned if necessary, and returned to the system.

A more specialized example is that offered by PPS for fish boxes (Figure 7.17). Under this system, fish caught by a trawler in Icelandic waters can be boxed on return to harbour, and then shipped by refrigerated transport to the UK. Here the fish can be sold and the box off-hired, to be cleaned and returned to Iceland. The advantages of being able to use a one-way hire service over such a long distance are very clear to see.

Labelling

When I was working for a 3PL we received a package addressed 'To Birmingham. For the attention of Dave.' When I called the sender he said, 'You have delivered to our Birmingham depot before, so I thought you knew where it was.'

For some reason, labelling is an often-forgotten part of the logistical process. The above is the most extreme example I can recall, but poor labelling is far from uncommon. Address labels should always include country, postal or zip code, and department or other identifier within the destination – the name of a person if that is all that is available.

Figure 7.17 Re-usable one-way hire boxes used for transport of fish, in this case cod from Iceland to the UK

SOURCE Reproduced by kind permission of PPS Equipment

A frequent mistake is to confuse delivery address with that of purchasing, accounts or head office. Whilst this can be due to simple carelessness, language barriers can create difficulties – the message should be 'If in doubt, ask.'

Always include an advice note. This can be placed within the package, or in a 'documents enclosed' envelope attached to the outside. It is best to include all the details you can think of – item number and descriptions, quantities, packages, and purchase order number – but not confidential information such as price. Major customers will probably detail their requirements, but it is far better to err on the side of caution and include more rather than less information.

Your customer may also specify certain information to appear on labels: for example, a TV shopping channel may wish their own unique item number to be shown. In such cases the label must obviously be visible. Another embarrassing mistake to avoid is to label the product, then place it inside an outer carton so that the label cannot be seen, or to cover the label with packing tape.

Barcodes

This is another item we all encounter every day, but which receive little thought. There are in fact numerous different barcode languages – Richards (2018) provides

Figure 7.18 Desktop barcode printer. Two barcodes have been produced, together with other conventionally written information

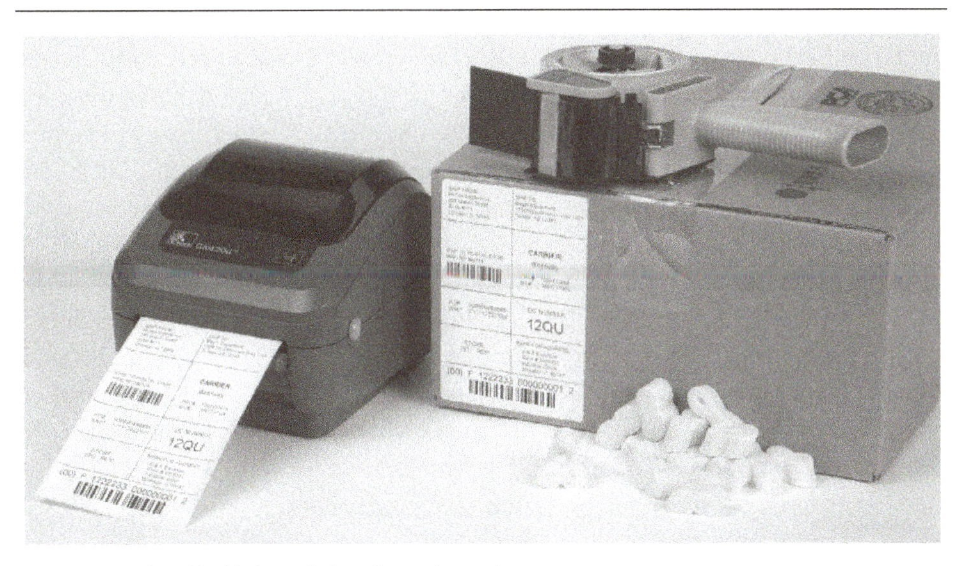

SOURCE Reproduced by kind permission of Barcode Warehouse

an appendix listing 15 of the more common ones. The most frequently seen in Western Europe is called EAN13 – if you are reading a printed copy of this book an example will be seen on the back cover. It is important to ensure that you use a barcode language that can be read by your customers' scanners – they will probably specify a choice of about three that you may use. It is also important to ensure that there is sufficient white space around the barcode, or it will not be readable. A former customer of mine had thousands of items rejected by a leading retailer for exactly this reason.

Free barcode generator software can be found on the internet, and this could be used to print barcodes onto labels using a standard office printer. However, I would recommend a purpose-designed printer to ensure that a high standard of printing is maintained. A desktop unit suitable for medium volume applications can be obtained for a few hundred pounds/dollars/euros (Figure 7.18). Printers can work on a stand-alone basis, or be incorporated into an automated process, the same unit printing and attaching the label as an item passes on a conveyor. If the unit can also print non-barcode information onto the same label, such as item number and description, this will be beneficial.

Customer requirements

I have already mentioned that major customers will probably have detailed specifications of the way in which they wish to receive deliveries: for example, Leidos (a defence contractor) supply an 86-page manual (Leidos, 2017). In the vast majority of cases, companies have good reason for doing so. They wish to receive all their purchases in the same format, which makes them easier to handle and saves wasted time – operatives searching for labels, for example. Indeed, stipulations are often pure common sense – broken pallets and obscured labels are understandably unwelcome.

Supplier requirements often extend far beyond logistics – multinational companies are likely to have policies that aim to eliminate modern slavery and child labour, and financial stability must often be demonstrated. I will, however, be restricting my discussion to logistics matters.

Having said that, it is sad to say that some customers have been known to use the small print as a cynical method of cancelling orders without incurring penalties. I recall a consignment of fans being rejected in a cold and wet summer on the grounds that the diameter of the wheels of the delivery vehicle fell outside the specified size range (in fact they were 2 cm within the range). My best advice is to ensure that you adhere rigidly to the requirements, and pre-agree any deviations.

Matters likely to be covered by customer requirements are:

- Pre-booking. There will almost certainly be a pre-booking system, which will require a minimum period of notice. If, for example, you have received an order from UK TV shopping channel QVC, you will have received a due date with your purchase order, and you should contact them for a booking at least 10 days before this date (QVC, 2018). It is wise to obtain confirmation of your booking if this is not issued automatically. Ensure that your vehicle arrives in good time, as they may have to queue, and there are unlikely to be resources available to unload if a vehicle arrives late. Failure to meet an appointment will probably result in the customer having under-utilized people and handling equipment and you may face a financial penalty – QVC currently charge £250 for missed delivery slots (QVC, 2018), and I understand their reasons for doing so.
- Creation of ASN. In some cases, for example Amazon, this may be required before the booking is made (Amazon, 2018); more commonly it is generated later. In all cases there are likely to be reference numbers which must be quoted – a bill of lading number in the case of Amazon, a booking reference number for QVC. An accurate figure for the size of the order (eg six pallets or 312 hanging garment sets) will also be required to enable resources to be planned, along with basic information such as items and quantities. Some customers will wish to be informed of the driver's name and vehicle registration number for security reasons (eg Leidos, 2017). If you are using a shared user network there may be a preferred supplier, such as Palletline in the case of Amazon.
- Creation of an advice note. The customer may specify a standard format for all their suppliers, and almost all will specify minimum information to be provided. Language is also important – in some countries such as the UK knowledge of

foreign languages can be poor and customers such as Selfridges will accept delivery documentation only in English (Selfridges, 2016).

- Labelling. Similar principles apply as for advice notes. Most customers will specify content and position required for labels, and barcodes in an acceptable language.
- Packing guidelines. Standard pallet sizes will be specified by most customers. This is likely to be a 1,200 x 1,000 millimetre base in the UK, and this is also a standard in NATO for the defence industry. Maximum heights and weights will, however, vary between customers, and in some cases between locations belonging to the same customer. For example, Wickes at Wakefield accept pallets up to 1 tonne and 1.5 metres high; Wickes at Nether Heyford up to 1.5 tonnes but only 1.3 metres high (Travis Perkins, 2018). I assume this reflects different racking and/or MHE at the two sites. Some may also specify closure methods – Selfridges (2016) do not accept steel or nylon banding for outer cartons. One well-known manufacturing organization insists upon clear tape only, and states that 'Coloured tape is unacceptable'. The rationale for this does I am afraid escape me.
- Presentation to retail customer. A few organizations require individual items to be arranged on a pallet or roll cage shelf, such that each faces the front, so that it makes an attractive display in the store.
- Vehicle requirements. Many warehouses are set up for unloading from the rear via dock levellers, and have no other facilities. This will mean restrictions relating to size of vehicle and minimum deck height. Also, if you have loaded from the side with two-way entry pallets, these will not be able to be unloaded from the rear. Conversely, if you are delivering to a site where unloading can only take place from the side, box van trailers with solid sides will be unsuitable.
- Arrival at site. Drivers should be briefed as to what is expected of them on arrival. High-security sites, such as those in the defence industry, may require photographic proof of identity (eg Leidos, 2017); drivers may require safety shoes and/or high-visibility clothing; passengers and animals may be banned from the site. The driver may be told to wait in a certain area, and keys may be taken away whilst the vehicle is being unloaded (this is a safety precaution to prevent it being driven away whilst a forklift is entering or exiting, which is a very dangerous occurrence). They may or may not be required to open and close doors and curtains, and/or to assist with unloading. In all cases the driver should be well briefed.
- Proof of delivery. Drivers should ensure they obtain proper proof of delivery, which meets customer stipulations. The customer may, for example, refuse to pay an invoice if the date and time stamped onto the delivery notes is incorrect, or if the stamp has been missed from one page, even if the total number of pallets delivered is not in dispute. Any alleged shortages, damages or discrepancies should be checked immediately – there may simply have been a miscount. Again, thorough briefing is the key.

- Fully or partially rejected consignments. Whilst this is something we all strive to avoid, it does happen. It is a good idea to have a contingency plan in place – if possible the same vehicle should bring back the unwanted goods. Failing that, it will probably be preferable to collect them at a later date: using the customer's carrier may well prove expensive.

Customer collection

In the above, I have made an implied assumption that you will be delivering to your customer. However, some customers insist on collecting, or offer it as an option. This is known as an ex-works system in the automotive industry, and as factory gate pricing in the retail sector. If your customer collects, almost all the above comments will still apply, and responsibility for transport is likely to be the only change.

Conclusion

In this chapter I have talked about how goods should be presented to the customer. This has included numerous options for packaging, labelling and the meeting of specific customer requirements. It should not be forgotten that each of these has a cost, and will be a contributory factor in the varying costs of serving different customers – a concept known as 'cost to serve'. There are of course other factors involved, some of which I have yet to mention, so I will be discussing cost to serve in more detail later in the book.

In the meantime, in the next few chapters I will describe various options for delivering goods either into your own premises or to the customer.

Road vehicles 08

At the beginning of this book I quoted a tongue-in-cheek definition of logistics as 'running sheds and lorries'. Whilst logistics is a much broader discipline, it cannot be denied that these are the two core functions. So far, I have mainly discussed what happens in the sheds, from what type of racking might be best for your requirements to appropriate packaging.

I will now move on to discussing the second main activity, that of transport. You may be tasked with moving goods into your company's sites; between them; or from those sites to your customers. You may be operating your own resources or using the services of a 3PL. However, in each case there is a strong possibility that you will be using road transport, especially for domestic movements.

In the UK in 2016 freight movements by heavy goods vehicles (HGVs) totalled 170 billion tonne kilometres, almost exactly 10 times the figure for rail movements of 17.05 billion tonne kilometres. In addition, there were 79 billion vehicle kilometres of van movements (FTA, 2018).

I will discuss rail, sea and airfreight later, but the first question to be asked regarding roadfreight is: 'What sort of vehicle should you be using?'

Two-wheeled transport

If I had been writing this book in the year 2000 and had mentioned pedal cycles, many people would have thought I was joking. However, for movements of small consignments over a few kilometres they are now a serious alternative. Cycling has become increasingly popular as a hobby and as a mode of personal transport, and there is a pool of keen fit cyclists available to work in the sector. Whilst perhaps most associated with take-away food deliveries, it is also a viable option for documents or pharmaceuticals, especially in inner city areas. The main advantages are low cost, environmental benefits, and the ability to avoid traffic congestion.

For very urgent movements over longer distances, same-day motorcycle couriers may be the best answer. These may include urgent medical supplies, spare parts for a broken-down vehicle, or promotional material required for a vital sales presentation. Costs in practice may be little if at all cheaper than for a small van, but there is again the advantage of avoiding traffic congestion, and it is often the most reliable method of delivering a small item quickly.

Panel vans

I was once asked 'How big does a van need to be before it becomes a truck?' There is no good answer to that question, but I can answer a slightly different one, and define a panel van – sometimes called an integral van. If the rear part of the vehicle has a roof but no windows and is designed for goods use, and the roof and sides are formed from panels continuous with the front of the vehicle, then it is a panel van. Smaller panel vans tend to be based on cars, using the same chassis and drive train and as many shared components as possible, such as the Vauxhall Corsavan. These are known as car-derived vans or CDVs. Larger vans such as the Mercedes Sprinter tend to be purpose-designed as vans. In the range between, vans can fall into either group – the Transit Connect is purpose-designed, but the Peugeot Partner of similar size is derived from the Peugeot 308 car.

All major manufacturers offer a range of vans in different sizes. In the UK, Ford market three of the top four best-selling vans: the other is the Volkswagen Transporter (January–July 2018, SMMT, 2018a). I will therefore use their models to illustrate the sizes available, and have also included the Mercedes Sprinter, which I believe is the largest panel van currently sold in the UK (Table 8.1).

I have used figures towards the top end of the range in each case. However, these may not be available together. For example, selecting a high-roof van will increase the space for freight, but the weight of the additional bodywork will reduce the maximum weight able to be carried.

At the smallest end of the market is the Ford Fiesta van (Figure 8.1) – competitors include the Vauxhall Corsavan. It is cheap both to buy and to operate, and easy to drive, especially around town; the main downside is obviously the small carrying capacity. The rear door will be a lifting tailgate rather than sideways opening doors, but this is unlikely to create many problems. However, there may well be a lip at the rear of the load compartment, and lifting heavy items over that lip might create a problem. Having said all this, if you are delivering small, light items such as flowers or cakes, especially on a local basis, then this might be all you need for that purpose.

If your goods are not as small, then a larger van will be required – although confusingly this may still be referred to as a 'small van'. Examples include the Ford Transit Courier (Figure 8.2), Transit Connect, VW Caddy, Citroen Berlingo and Peugeot Partner: the last two are very similar models from sister companies in the PSA Group. Such vans will have about three or four times the capacity of a Fiesta

Table 8.1 Sizes of a range of Ford vans plus a Mercedes Sprinter

Model	Gross vehicle weight (kg)	Overall length (m)	Load length (m)	Load space (m^3)	Payload (kg)
Fiesta van	1,685	3.95	1.30	0.96	500
Transit Courier	1,795	4.16	1.60	2.40	663
Transit Connect	2,395	4.82	2.20	3.60	865
Transit Custom	3,365	5.34	2.90	8.30	1,393
Transit	4,700	6.70	4.20	15.10	1,446
Mercedes Sprinter	5,500	7.37	4.70	15.50	2,343
(5.5 tonne option pack)					

SOURCE Manufacturers' brochures: Ford (2018) and Mercedes (2018)
NOTE Figures quoted are each the highest for the range stated, and particular combination may not be available

Figure 8.1 Two Ford Fiesta vans

SOURCE Reproduced by kind permission of Ford Motor Company Limited

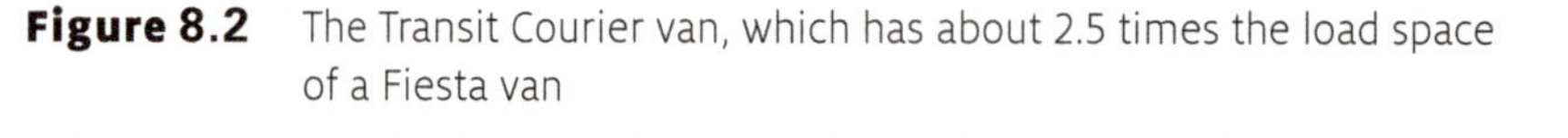

Figure 8.2 The Transit Courier van, which has about 2.5 times the load space of a Fiesta van

SOURCE Reproduced by kind permission of Ford Motor Company Limited

van, but still retain the drivability of a medium-sized car. Access will be via sideways opening rear doors, and there is unlikely to be a lip, which makes handling of heavy items much easier.

At the larger end of the market comes the Transit and its competitors. The name Transit has been in use since 1965 (Transit Owners' Club, 2018) and is often used as a generic term for all vans in the class, a little like using the name Hoover to describe any upright vacuum cleaner. The current Transit Mark 8 is of course very different to the Mark 1 of 1965. Competitors include the VW Transporter, Peugeot Boxer, Mercedes Sprinter and Nissan NV400. These vehicles drive differently to cars, and whilst most drivers will be able to handle them without undue difficulty, awareness of the size of the vehicle is necessary. It should not therefore be assumed that anyone with a car driver's licence is competent to drive a large van.

Factors to consider when selecting a van include the following.

Size

Almost all of the models I mention come with a bewildering range of options. Unlike the car market, this includes options for size. Variants such as short, long and extra long; or low, mid and high roof, affect the available load space in terms of cubic metres; and different weight ratings affect the payload in terms of kilogrammes. Variants of the Nissan NV400, for example, range from 2.6 to 4.4 metres cargo length; 911 to 2200 kilogramme payload; and 8 to 17 cubic metres (Nissan, 2018). If your product is heavy, such as wheel bearings, a smaller cube but high payload is desirable. To opt for a long-wheelbase, high-roof vehicle in which the cube will never be used will be counter-productive, as the weight of the extra bodywork will reduce the deadweight capacity.

One pitfall to avoid is to look solely for the highest payload in the brochure. This will probably be a variant that gains a few kilogrammes of payload due to a smaller

Figure 8.3a–b Two different models from the Ford Transit range, showing a clear difference in length of load compartments

SOURCE Reproduced by kind permission of Ford Motor Company Limited

engine, and it may well prove under-powered and sluggish in service. In reality, most potential customers test drive vans empty, but they handle differently with a full load, so if you can arrange a loaded test drive I would recommend doing so. Similarly, good headline fuel consumption will not be realized in practice if a small engine necessitates over-use of lower gears.

If you are carrying very long items, such as TV aerial poles, some models offer the option of folding down the front passenger seat to give additional load length.

> Vans above 3.5 tonnes, including the weight of any trailer, are legally treated as HGVs. You will need an operator's licence, drivers will need to use tachographs and obey drivers' hours regulations, and unless they passed their test before 1997 people will not be able to drive them on a standard car licence. Unless you are prepared to take on these serious legal responsibilities you should limit your fleet to vehicles below this size, or use a contractor.

It should be remembered that a larger vehicle will cost more to buy; road tax, fuel costs, maintenance, tyres and insurance will be more expensive; it will be more difficult to drive and manoeuvre in tight spaces; and will be slower and therefore less productive in service. My generalized advice would therefore be to calculate the size of vehicle you need, now and in the future, add a small margin for contingencies, and buy a vehicle large enough to meet those needs, but no larger.

Access doors

Apart from the smallest vans, a pair of sideways opening rear doors will probably come as standard. On some models asymmetric doors are available, ie one door is wider than the other (Figure 8.4). This can be useful if carrying awkward items.

A side door is also commonly seen, and is generally a good idea (Figure 8.5). It allows access to the load compartment from more than one angle and makes handling a lot easier, especially if collecting and delivering on the same route. It is difficult to think of circumstances in which I would not recommend at least one side door, and some vans have two.

Transmission

Most vans come with the option of manual or automatic transmission. Most drivers in Europe are more familiar with manuals (the opposite applies in the United States) and I would prefer a driver to be in full control and able to select lower gears at will

Figure 8.4 Transit Courier van, showing asymmetric doors

SOURCE Reproduced by kind permission of Ford Motor Company Limited

Figure 8.5 Rear and side doors in use on Transit Custom van

SOURCE Reproduced by kind permission of Ford Motor Company Limited

if tackling an incline with a heavy load. Fuel consumption is likely to be marginally better with a manual, and a cost difference of over £1,000 for an automatic Transit versus an equivalent manual version is not insignificant. Some will disagree with me, but my own preference would be for manual transmission.

Drive

There are three options – front wheel drive (FWD), rear wheel drive (RWD) and all wheel drive (AWD).

In most cases, my recommendation would be for FWD. This avoids the need for the prop shaft and complex rear axle to be under the load compartment, saving space and weight. This will give better load space and greater payload, and the loading deck will be lower, which makes lifting heavy objects in and out of the van easier. FWD are cheaper to buy than RWD equivalents, and give better fuel consumption.

RWD is the better option if towing a trailer – the driven wheels being closer to the towing point gives an advantage when doing so. Turning circles will be tighter, which may be important to some operators, and with some model ranges there is a wider choice of engine offered for RWD vehicles.

AWD is more expensive than either option. It is better if it can be engaged or disengaged as required, as this will improve tyre wear. If you are likely to be operating in slippery conditions – for example muddy construction sites or icy mountainous areas – then AWD will clearly be advantageous. If you are operating solely on tarmac roads in a warm climate, the extra cost is unlikely to be justified.

Driving aids and safety features

A very small proportion of van drivers (dubbed 'white van man' by the popular press) sadly give their profession a bad name. They have a reputation for driving too aggressively, and appearing to think that the size of the vehicle gives them extra protection in accidents, rather than being a reason to drive more carefully to avoid accidents.

However, the vast majority of van drivers, in my experience at least, fit a very different profile. They are highly professional and take great pride in their vehicles. Good personnel management, such as providing relevant training and monitoring accident damage, plays an important role in maintaining and improving driving standards.

Additionally, there are optional features available that help the driver to maintain high standards, monitor performance, or both. These include:

- Parking sensors. Give both visual and audible warning as the vehicle approaches an object such as another vehicle.
- Rear-facing camera, so that the driver can get a good view on a screen when reversing.
- Active park assist. Takes this one step further by taking control of the van to manoeuvre it into a parking space.
- Speed limiters, often set to prevent a driver exceeding the national speed limit for the particular vehicle.

- Hill start assist. Helps to prevent the van rolling backwards or stalling the engine when starting facing uphill.
- Traction control. Helps to prevent wheels slipping on mud or ice.
- Cruise control. Often useful in avoiding speeding prosecutions when the speed limit is lower than it could be, for example in road works
- Lane keeping assist. Designed for motorways, a forward-facing camera activates a visual alert and vibrates the steering wheel if the van veers out of lane, and if the driver does not react it can provide steering assistance to help guide the vehicle back into lane.
- Traffic sign recognition. Recognizes signs such as speed limits and 'no overtaking', and displays an icon on the instrument panel accordingly.
- Attention assist. Detects signs of drowsiness and tells the driver to take a break.
- Dashboard cameras. Help to identify the party at fault in case of accident, so that action can be taken against a driver, or a spurious claim from another motorist rejected.
- Pre-collision assist. Detects pedestrians and vehicles in front, and sounds alarm and/or applies brakes if necessary.
- Automatic headlights, which activate in poor light conditions.
- Tyre pressure monitors, which can be viewed from inside the cab.
- Trailer sway control. Reduces engine torque and applies brakes to one side of towing vehicle if sway is detected.
- Emergency assistance. Automatically notifies a control centre if the van is involved in an accident that is serious enough to cause an airbag to activate. It works in 40 European countries, and is designed to avoid situations in which an injured driver is trapped in a vehicle for hours or even days before being found.
- Satellite navigation. Can be voice-controlled to enable driver to use both hands on the wheel.
- Additional security. There are innumerable possibilities for enhanced alarm systems, high-security locks and tracking devices. Whilst much can be achieved by good discipline, such as not leaving vehicles unattended, such products can play a useful role. For very high-risk cargoes such as cash, armoured glass and bullet-proof fuel tanks should be considered.
- Incorrect fuel-fill prevention. Draining a fuel tank if filled with petrol rather than diesel is expensive, and will result in the vehicle being out of action for several hours. If petrol actually gets into a diesel engine the consequences are even more expensive. I would recommend devices to prevent this, either permanently fitted to the van or as accessories.

To fit all of these devices to a single vehicle would cost a great deal of money, and it is unlikely that anyone has ever done so. Hill start assist will be of little use in Flanders or the English Fens, and lane keeping assist of no benefit to those whose driving is restricted to inner cities. However, it is worth considering each option on its merits for your particular needs, and choosing accordingly.

Driver and passenger options

There was a time when van interiors were very basic, not a pleasant place to spend one's working life, and a serious cause of driver fatigue. However, modern vans usually provide car-like comfort in the cab as standard so this is rarely an issue (Figure 8.6). One optional extra worth considering is heated seats. If drivers are making night-time deliveries in sub-zero winter weather, which results in them getting in and out of the cab numerous times during their shift, this will be most welcome.

Most vans have a driver's seat and a single passenger seat. However, double cabs with two rows of seats (usually five in all) are available. This can be useful if a work team and their equipment need to be taken to site, for example in the construction industry. A third option is the Transit Kombi and its competitors, in which the second row of seats can be folded flat when not in use, which is a more flexible alternative.

In-cab storage is useful, for example a drawer under the passenger seat. This can be used for the driver's personal effects and/or paperwork for the deliveries being made. It is best to keep such items out of the main load area.

Figure 8.6 Inside the cab of a modern van. This is almost indistinguishable from a car

SOURCE Reproduced by kind permission of Ford Motor Company Limited

Load area options

A hardwearing floor covering to the main load area will come as standard. I would also recommend plywood or other lining to the sides to prevent bodywork damage, which would otherwise be inevitable over the life of the vehicle. Mesh covers for the rear and/or bulkhead windows will also help prevent damage (Figure 8.7).

Alternatives are often offered for the front bulkhead. I would recommend a mainly solid bulkhead, as clearly visible items in the van, even boxes of unknown contents, tend to have an attraction for opportunist thieves.

Ample tie-points for load securing straps are essential. These may be floor-mounted, side-mounted (perhaps spaced along a waist-bar), or both.

The commonest scenario is that of vehicles being used for collections and deliveries, with material loaded on the floor. However, sometimes it is not known what will be required when reaching the destination, for example if goods are being sold from the van, or a roadside assistance vehicle is attending a breakdown. In these cases, some form of racking or shelving in the load area will be very useful. Simple racking is often available from the vehicle manufacturer, but it may be better to use a specialist to fabricate racking optimized for your particular needs.

If personnel will be entering the load compartment, steps up from ground level should be provided.

Figure 8.7 Interior of a van's load compartment

Plywood lining to walls; tie-down points just above floor level and on top of the wheel arches; protective mesh over bulkhead window; two side doors; and steps up from ground level inside side doors

SOURCE Reproduced by kind permission of Ford Motor Company Limited

A roof rack is a common addition, useful for long items such as a ladder that will not fit into the load compartment. If using a roof rack, steps should be provided to enable the driver to reach it safely. An unattended ladder on a roof rack can be at risk of theft, and it is possible to fit a ladder rack on the underside of the roof (ie inside the van) to prevent this.

Finally, the load compartment should be well lit. It may also be useful to fit an outward facing light by the rear door if night-time deliveries to unlit locations are being made.

Electric vans

There is a general trend towards rechargeable electric vehicles, and as technology advances they will undoubtedly become more common. Mercedes already market the eVito van. This is a medium sized van with 6.6 cubic metres load space and just over one tonne of payload. Charging takes six hours, and is sufficient for a range of 93 miles. An eSprinter version of the larger Sprinter van is scheduled for launch in 2020 (Mercedes, 2018).

These are, I am sure, a sign of things to come.

Pick-ups

The integral pick-up differs from an integral (or panel) van in that the load compartment is typically only about half to two-thirds the height of the vehicle's body, without a roof, and with a drop-down tailgate at the rear.

They are widely seen in the USA, accounting for 16.5 per cent of car sales (Business Insider, 2018), and often used as the main family vehicle for domestic purposes. However, they are much less common in Europe, representing just 1.8 per cent of car and van sales in the UK (January–July 2018, my calculation based on SMMT, 2018b). Some are no doubt purchased for domestic use, but as commercial vehicles their main role is in the agricultural and construction sectors. Models available include the Ford Ranger (Figure 8.8), Isuzu D-Max, Toyota Hilux and Nissan Navara. Options available include AWD, which is especially useful in agricultural environments.

Whilst a cover (called a tonneau cover) can be used to protect goods in the load compartment, it will not give the same level of protection from the elements as a van. Open-top alternatives such as dropside vehicles are available, and a pick-up would rarely be my recommended solution.

Figure 8.8 Ford Ranger pick-up in use on a farm

SOURCE Reproduced by kind permission of Ford Motor Company Limited

Larger vehicles

Some aspects of describing possible types of vehicle can be extremely complicated – the legal regulations regarding vehicle size are confusing enough in themselves.

In the simplest terms:

- A vehicle can be a rigid of variable size, an artic, or a rigid and trailer combination.
- Various configuration options are available for the vehicle.
- The load compartment can be a body of different types, some more specialized than others. It is not much of an exaggeration to say that any body type can be fitted to any size of vehicle.

I will discuss each of these in turn.

Other than integral vans and articulated vehicles, most rigid commercial vehicles seen on the road will have been manufactured as a chassis cab, with bodywork added later, probably by a specialist bodybuilder. They can vary in size from a Nissan NT400 with a 2.8 tonne gross vehicle weight (GVW) (Nissan, 2018) to a Volvo FMX with 4 axles and a 36 tonne GVW (Volvo, 2018).

Most 3.5 tonne GVW vans are alternatively available as a chassis cab, and these have the advantage that they require no operator's licence and can be driven on a standard driving licence. Also frequently seen are 7.5t GVW rigids – older drivers can drive these on their standard car licence (see Chapter 9 for further details). Another common size is the 18-tonne rigid, which is the maximum permitted size of a two-axle rigid (Lowe and Pidgeon, 2018), and the most common vehicle for multi-drop pallet deliveries. The largest rigids include tipper vehicles seen on motorway construction sites.

The most visible parts of a chassis cab are just those two elements. There will typically be a cab with seats for driver and passenger, with a panel immediately behind. There will be a basic chassis, usually with two substantial longitudinal beams and one or more cross beams. The engine, gearbox, wheels, brakes and all other mechanical parts for the running of the vehicle are also provided, but there is no load compartment: this will be added later. A chassis cab pictured before fitment of a body is shown in Figure 8.8, and the same model of chassis cab with body fitted is shown in Figure 8.19.

An articulated combination – almost always known as an artic – consists of a tractor unit (also called a prime mover) and what is technically called a semi-trailer, but in reality is known to almost everyone simply as a trailer. A kingpin on the underside of the trailer will engage with the fifth wheel on the tractor unit to form the coupling (Figure 8.10).

The maximum length of an artic trailer in the UK is an example of what many would consider unnecessarily complex rules. The maximum length from the kingpin

Figure 8.9 Isuzu N Series chassis cab

SOURCE Reproduced by kind permission of Isuzu Truck UK. Photography by Nigel Spreadbury

Figure 8.10 A tractor unit manoeuvring to couple with a trailer

SOURCE Reproduced by kind permission of LC Vehicle Rental

to the back of the trailer is 12 metres. The maximum dimension measured forward from the kingpin is 2.04 metres, which is measured diagonally to the front corner of the trailer (Lowe and Pidgeon, 2018). In practice, this gives a maximum of 13.6 metres loading length – although there are exceptions, eg for car transporters. Maximum GVW for an artic is 44 tonnes, subject to various stipulations including engine emission standards.

A greater loading length can be obtained using a rigid vehicle hauling a drawbar trailer – this is also known as a roadtrain or wagon and drag (Figure 8.11). The main advantage is that a combined loading length of 15.65 metres is permitted, ie about 2 metres or four 1,200 x 1,000 millimetre pallet spaces more than an artic in the UK. In most countries in continental Europe, trailers are limited to 12 metres in length, so the difference is greater, and this is even more important as there is a 4 metre height limit in many European countries. As a consequence, rigid and drawbar combinations are more commonly seen in continental Europe than in the UK, where they remain unfashionable. Maximum GVW is 36 tonnes (Lowe and Pidgeon, 2018) but most loads tend to reach the length and/or volume weight of a vehicle before they reach the weight limit. For a very light product such as breakfast cereal maximum height variants of such vehicles have a clear advantage. Disadvantages include the need to divide the combination before unloading from the rear, and that it is not possible to operate a drop trailer system. Also, UK drivers may be unfamiliar with such vehicles, although it is perhaps surprising that they are not more common in the UK.

Figure 8.11 Two rigid and drawbar combinations entering and leaving a depot

SOURCE Reproduced by kind permission of Eddie Stobart Logistics

Vehicle and engine size

There is a tendency to source vehicles at the maximum GVW for their class. This has the advantage of flexibility if the business should change, greater availability on the 3PL market if you are using a contractor, and likely higher residual value than for the less popular types. However, vehicles at GVWs other than the maximum for their class will be cheaper to buy and to operate. For example, a 13 tonne rigid will typically have annual standing costs of £55,620 and running costs of 42.4 pence per mile (ppm), compared to £64,191 and 47.0 ppm for an 18 tonner (Motor Transport, 2018b) – a saving of over £10,000 per year based on 50,000 miles. If you have a light product such as plastics and can therefore be confident that such a vehicle will meet your needs this will be a worthwhile saving.

Regarding engine size, the best advice is to discuss options for your particular needs with your vehicle supplier. However, it is important to ensure that the vehicle is not under-powered: no one wants to see a return to vehicles crawling up hills on motorways at 30 mph. This not only gets the industry a bad name but is also unproductive, resulting in a vehicle making insufficient deliveries in a given time period. A margin must also be allowed for power take-off for ancillary equipment such as refrigeration units or a cement mixer, where this is envisaged.

Arduous use, such as running mainly with full loads and/or in mountainous areas, is another consideration. Taking all this into account, for 6 x 2 artics (see the following box for an explanation of this term) 450 to 520 horsepower engines are the most popular as of 2018. However, for refrigerated operations 600–650 horsepower may be more appropriate.

Axle configurations and vehicle lengths

For artic units and rigids, the standard nomenclature for axles is in the form A x B, where A is the total number of wheels, and B is the number of these which are powered or driving wheels. If there are two tyres on a wheel this still counts only as one wheel. Therefore a 6 x 2 rigid, for example, will have six wheels (three axles), of which two (one axle) are driving wheels; an 8 x 4 will have eight wheels (four axles), of which four are driving wheels.

For trailers, it is the axles that are counted. An artic trailer might be single axled (rarely seen these days); tandem axled, ie with two close together; or tri-axled.

There are three main considerations when deciding on axle configurations for your vehicles – gross vehicle weights, axle weights and bogie weights for axles close together. To operate legally, a vehicle must remain within the specified limits for all of these, bearing in mind that the plated limit (ie that for the particular vehicle) may be less than the legal maximum. For example, a 4 x 2 artic hauling a tandem axle trailer may legally have a maximum GVW of 38 tonnes (subject to restrictions, Lowe and Pidgeon, 2018), but a single heavy item such as an engine loaded at the front of the trailer will probably cause the axle weight on the rear axle of the tractor unit to exceed the legal limit of 11.5 tonnes for a single drive axle. A tandem pair of driving axles would have a combined maximum weight of 19 tonnes (again subject to restrictions, Lowe and Pidgeon, 2018), and a 6 x 4 tractor unit would therefore avoid this problem. Whilst it may appear counter-intuitive, it is possible for all axle weights to be within legal limits when the vehicle is fully loaded, but to become illegal when the rear part of the vehicle but not the front section has been unloaded. Additional axles reduce the risk of exceeding axle and bogie weights and therefore improve flexibility of loading, and this is one reason why four-wheeled tractor units represented just 8.4 per cent of UK artic registrations in the first half of 2018 (author calculation based on SMMT, 2018c).

Exceeding the maximum permitted weights is a serious offence, and there is no defence of due diligence: if your supplier loads more freight on your vehicle than they have stated such that it becomes overloaded, this will not excuse you or your driver. The only defence is if you can prove you were en route to the nearest weighbridge. Most manufacturers now offer in-cab instruments to indicate loading weights, some of which can be monitored remotely and even by mobile phone, and I strongly recommend their use.

> If you are tempted to under-declare the weight of your goods when loading a 3PL vehicle in the hope of obtaining a cheaper freight rate, this is not only a fraud, but is dangerous, and could result in the vehicle driver being prosecuted and losing his job and his livelihood. Do not do it.

Whilst it is inefficient to operate empty vehicles, it is in some cases unavoidable. In these circumstances additional axles are unnecessary, and many vehicles are therefore equipped with lifting axles. The driver is able to activate this function from the cab, and one of the axles on the tractor unit and/or trailer can be lifted clear of the road surface by air, electric motors or springs (Figure 8.13). This reduces fuel consumption by up to 4 per cent (Volvo, 2018) and reduces tyre wear. Some operators have been known to lift axles when the vehicle is fully loaded, but this is a false economy as it results in excessive wear on the non-lifted axles. Some US states, eg Georgia, have declared that vehicles must meet legal limits at fully loaded weight even when all lifting axles are lifted, in an attempt to eradicate this practice (Trucking Info, 2010).

With regard to vehicle lengths, the majority of vehicles built to the maximum GVW for their class also tend to be built to the maximum permitted length. Most rigid vehicles at 18 tonne GVW are 12 metres long overall, and most artics are 16.5 metres in total, to maximize the body length available for loading. It is possible to reduce body length to decrease the tare (empty) weight of the vehicle and thereby increase payload, a tactic sometimes followed to keep within the 7.5 tonne GVW limit. However, this reduces the flexibility for distributing a load within the vehicle

Figure 8.12 Load monitoring for axle, bogie and gross vehicle weight

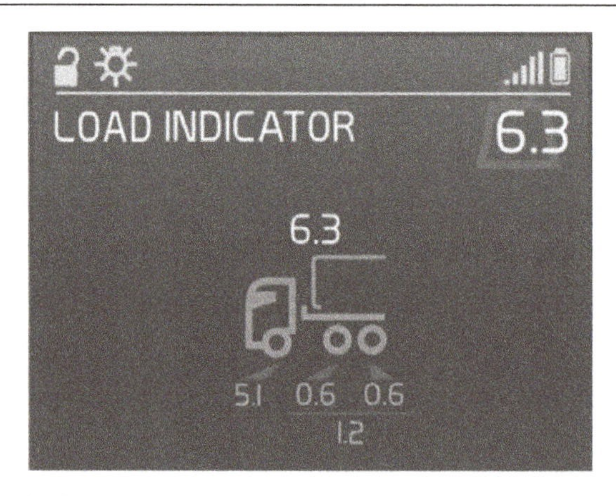

SOURCE Reproduced by kind permission of Volvo Trucks UK and Ireland

Figure 8.13 Vehicle with lifted axles on both tractor unit and trailer

SOURCE Reproduced by kind permission of Volvo Trucks UK and Ireland

and for carrying larger loads, and the residual value will be significantly reduced if a vehicle is later sold; lease companies may be reluctant to provide such vehicles for this reason. Unless there are unusual circumstances (for example, one company which I worked with operated 9 metre trailers due to space limits and access restrictions at very old brewery premises), the advantages of a longer body will tend to outweigh the disadvantages.

Cabs

Smaller vehicles are most likely to be making multiple deliveries during the working day, and returning to their base each night. The same is true of some artic operations, including those that are used by different drivers on day and night shifts. The appropriate cab in such cases is known as a day cab, with seats for driver and passenger but no sleeping accommodation.

Crew cabs are also available, with a second row of seats, so that a crew of perhaps six people including the driver can be carried. There are a few cases in which it is important for a large team to travel with the vehicle – a fire engine would be one example – but these are the exception rather than the rule.

If a driver remains in the cab overnight, then a sleeper cab will be needed. This includes a bunk behind the seats to enable them to get a good night's sleep.

A well-refreshed driver will be much more productive, and drive more safely, the following morning. For this reason, when a choice of bed width or comfort level is offered I would not favour the cheapest option – a single minor accident will cost much more than the money superficially saved.

Indeed, some companies provide a bunk even if overnight use in not anticipated (Figure 8.14). It provides a capability if circumstances change; a driver may wish to take a short nap whilst taking a driving break; and the residual value of artics with sleeper cabs is greater if the vehicle is sold.

For some very long-distance work two drivers need to travel together. Double sleeper cabs, with two bunks, are the solution.

In all sleeper cabs, ample storage should be provided for the driver's personal possessions. Curtains or blinds are another standard feature, as is a night heater and/or cooling unit to maintain a comfortable temperature. Heavy-duty heaters are an option for use in very cold climates such as northern Scandinavia. I would recommend a separate battery circuit for cab heating – to be unable to start the engine because all battery power has been used overnight is an embarrassing situation, to say the least.

Other optional extras for driver comfort include microwaves, fridges, coffee makers, TV points and Wi-Fi connections. The latter may be particularly welcome to allow a driver to contact their family by Skype whilst away from home. A drying unit for wet clothes will also be very welcome in some circumstances.

Figure 8.14 Bunk in a sleeper cab

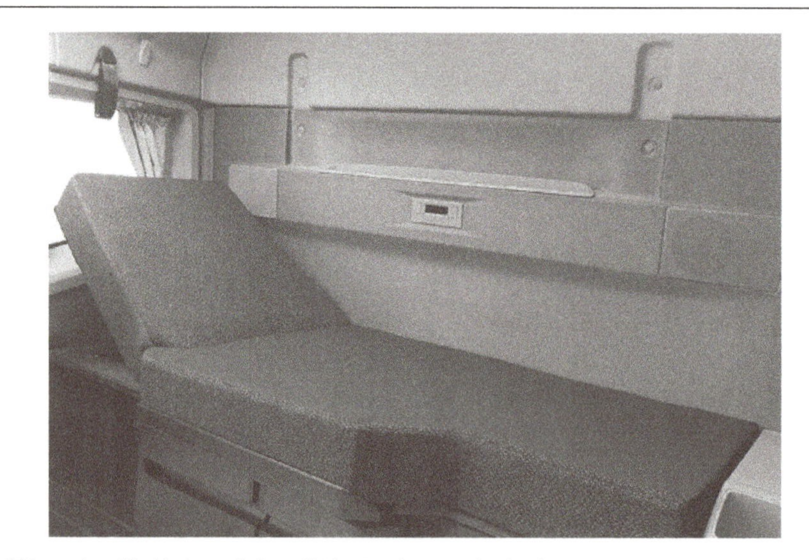

SOURCE Reproduced by kind permission of Volvo Trucks UK and Ireland

Air management

Aeronautical engineers have long known the importance of aerodynamic streamlining to reduce drag caused by the shape of the aircraft – known in that industry as 'form drag'. The same principles are now applied to trucks. Artic units can be fitted with roof, side and bumper spoilers, and side deflectors to minimize air resistance around the tractor unit and between cab and trailer. This can reduce fuel consumption by up to 10 per cent (Volvo, 2018). Trailers are now seen with teardrop-shaped roofs for similar reasons, and the next development may be 'boat tail' deflectors on the rear of trailers to reduce drag caused by turbulence immediately behind the vehicle.

Fuel tanks and AdBlue

Fuel tanks are available in a wide variety of capacities, varying from 150 to 900 litres for the Volvo FH16 for example (Volvo, 2018). Carrying almost a tonne of fuel will have a significant effect on the payload that can be carried. However, to select a particularly small tank in the pursuit of absolute maximum payload will require frequent stops for refuelling, which will reduce productivity and may occur at a very inconvenient time. There is no right answer to suit all eventualities, but I would say that any tank that requires a fuel stop more than once per day is too small. Other factors to be taken into consideration are distances between available fuel stops, which may be limited by factors such as fuel card networks (see Chapter 9), especially if operating at night when many fuel outlets will be closed.

Locking fuel caps and anti-siphon devices are options to reduce the risk of theft from diesel tanks, and are a sensible precaution.

AdBlue is a trade name for diesel exhaust fluid. It is composed of urea and demineralized water. It is injected into the exhaust fumes of a diesel engine, and reduces emissions of nitrogen oxides. It is kept in a separate tank on the vehicle and Motor Transport (2018b) estimate usage at about 4 per cent of fuel usage. It is not yet in widespread use in cars and vans but this may become compulsory in the future.

Fifth wheels

Most artics will be fitted with an adjustable fifth wheel. It is possible to fit a fixed fifth wheel, which will be cheaper, but this is likely to prove a false economy. Self-lubricating fifth wheels are also recommended, as is an in-cab indicator to confirm trailer attachment.

Driving aids

Many of the optional features listed earlier in the section on vans are also available for trucks – hill start assist, lane assist, driver attention devices, etc. Other options available from some manufacturers include:

- Dynamic steering. This applies additional steering power at low speeds, keeps the steering wheel straight if braking under difficult braking conditions, and reduces steering kick-backs on rough terrain.
- Advanced braking systems, which reduce the risk of jackknifing and other accidents.
- Adaptive cruise control, which helps the driver to keep a safe distance behind a vehicle.
- Hydraulic retarders. These operate on the prop shaft, using hydraulic fluid to provide an additional braking effect.
- Brake lining warning. These show the distance remaining before brake pads need replacement.
- I-shift. A system marketed by Volvo which ensures that gear changes take place at the optimum point (Volvo, 2018). Other manufacturers may offer similar systems.
- Alcohol locks. These require a driver to provide a breath sample before starting the engine, which will be locked if the sample reveals the driver has been drinking alcohol.
- Rough terrain options, such as heavy-duty bumpers, oil sump guards, light protectors and additional air filters.
- Remote monitoring. Again, these can be accessed by mobile phone applications. They can be used to remotely check on fuel, oil and battery levels; to schedule or activate cab heating systems; and to alert the driver via a smartphone if the truck's alarm system is activated.
- New driver apps. These help a driver get the best from a new vehicle, with short informative films on matters such as how to adjust driving positions. They are often free.

Finally, a driver should be provided with ample lights. This should include the area around the back of an artic cab to aid attaching Suzies (leads from unit to trailer).

Body types

Flatbeds

The simplest type of 'body' is the flatbed – a simple platform with a headboard at the front. An example was shown in Figure 2.3 (page 000). Their main advantage is ease of loading – there are no obstructions in any direction – and if loading and unloading is from above using a crane this will be the quickest type of vehicle to load. They are therefore often used in the construction sector, for example to transport concrete blocks or rectangular steel joists.

Their main disadvantages are the need to secure a load, and lack of protection. It is quite rightly an offence to drive a vehicle with an insecure load (not to mention the consequences of losing the load) and any item on a flatbed should therefore be secured with ropes, straps, or chains. A tarpaulin sheet may also be necessary. These can weigh up to 50 kilogrammes, and sheeting a trailer is therefore a physically demanding task. Unless the load is intended for outdoor use or not vulnerable to damage, such sheets will also be required to protect it. Unfortunately, water tends to find a way of penetrating even the smallest gaps in covers, with the attendant risk of damage to the cargo. This is one of the reasons they are less commonly used than in the past, and I would not recommend the use of an open vehicle to transport any product that requires protection.

Posts along the sides of a flatbed are sometimes seen, to help secure the load, for example when transporting large timber logs (Figure 8.15). A tailboard can also be provided, for example on trailers designed as bale carriers in the agricultural sector.

A specialized variant is the coil carrier, with one or more curved recesses in the bed to accommodate coils of steel.

Pick-ups, dropsided vehicles and tippers

As described above, integral pick-up vehicles represent over 16 per cent of car sales in the USA. However, it is also possible to add a pick-up body to a chassis cab, with

Figure 8.15 Flatbed vehicle with tall side posts being used to transport timber

SOURCE Reproduced by kind permission of Annandale Transport Co Ltd

fixed sides and a drop-down tailgate at the rear. This will in practice give a slightly wider load space with vertical rather than shaped sides.

More commonly seen are dropsided vehicles on which, as the name implies, the sides can be folded down when loading and unloading. They are commonly used in the construction industry, in all sizes from 3.5 to 36 tonnes, for the transport of building supplies, plant or both (Figure 8.16). Usually the sides are solid and less than a metre high, but sometimes they are taller and of mesh construction, for example on vehicles used for waste paper collections. Another application is for commodities that must be carried on open vehicles by law, such as propane gas cylinders (to prevent a build-up of explosive gases inside a closed vehicle). The ability to load and unload from the side gives greater flexibility than a pick-up, and given the choice would almost always be my personal preference.

The other type of open vehicle commonly seen is the tipper (Figure 8.17). Again, they are commonly used in both the construction and agricultural sectors, and the body is usually shaped to aid offloading when tipped. The nature of their work with loads such as aggregates and spoil is arduous, to say the least, and bodywork is usually of a very solid construction to facilitate this. Bulk tippers have high sides, and are used for example for scrap metal movements.

A roll-back cover is often provided, more with the intention of preventing loose stones falling off the vehicle and damaging others rather than protecting the load. Tippers designed for higher value goods – grain for example – will be fitted with a more substantial roof, and indeed would not be considered an open vehicle.

Figure 8.16 Ford Transit 3.5 tonne chassis cab with dropside body fitted, in use on a construction site

SOURCE Reproduced by kind permission of Ford Motor Company Limited

Figure 8.17 Tipper vehicle being loaded with aggregates

SOURCE Reproduced by kind permission of Avon Materials Supplies

The tailgate can be manual, opened by gravity when the vehicle is tipped; electric; or hydraulic. Grain tippers usually have a purpose-designed unloading chute.

Curtainsiders and box vans

Probably the most common type of body, either fitted to a chassis cab or on an artic trailer, is the curtainsider – also known as a tautliner (Figure 8.18).

Curtains are usually made from a polyester weave with a PVC coating, and designed to be load-restraining, so that if a pallet shifts in transit it will not be a danger to other road users. Curtains require tensioning before the vehicle departs, hence the alternative name. Usually there are end poles at front and rear on either side with tensioner screw gear, and two handles at the bottom – one to tighten and one to release. Straps at the bottom of the curtain are tightened with a ratchet and secured to a lip below the floor of the trailer.

Useful hints for inexperienced operators are to wear thick gloves; to use your leg muscles and not your back muscles when pulling the curtain, to reduce the risk of back injury; and to tuck in loose straps to prevent them flapping whilst the vehicle is moving.

Figure 8.18 Artic unit and curtainsided trailer – a very common sight on the roads

SOURCE Reproduced by kind permission of Denby Transport Limited

The main advantage of a curtainsider is obviously that the curtains can be opened to permit loading and unloading from the side. This process takes time, and if delivering a large number of small parcels then a van with a side door or doors will permit more deliveries in a given time. Whilst 3.5 tonne curtainsiders are therefore much less common than vans of the same size, they are useful if goods are delivered in small quantities but are difficult to manoeuvre through a door – a supplier of garden benches is one user who springs to mind.

For larger quantities of palletized goods and/or large and awkward parcels, which require side access for loading or unloading, the curtainsider is the standard solution, whether on a rigid or artic.

The disadvantages of side curtains are that they can be damaged, perhaps by forklifts when loading from the rear, and they require maintenance, such as greasing of screw gear. One option is to fit vertical poles inside the curtains with brackets to support wooden planks. Removing and replacing the planks to allow side access to the load is time-consuming and this is not an option I would recommend if frequent side access were required. However, if loading is usually from the rear (or from above, as in Figure 8.20) they may offer useful protection. Slashing of curtains, perhaps to see what is inside the trailer with a view to theft, does unfortunately happen.

If it is known that vehicles will always load and unload from the rear, a box van will probably be the best solution (Figure 8.19). This is a trailer or rigid vehicle with no side access, and doors only at the rear. The metal sides will be less prone to damage and more secure, and require no maintenance. These are used, for example, by

Figure 8.19 Isuzu N Series 7.5 tonne box van. This is built on a chassis cab as shown in Figure 8.9

SOURCE Reproduced by kind permission of Isuzu Truck UK. Photography by Nigel Waller

major retailers for trunking between their own depots, when it is known that side access will never be needed.

Whilst most curtainsiders and box vans have a solid roof, this will clearly preclude loading vertically from above. Some products must be loaded in this way, but require the protection of an enclosed vehicle – a large diesel engine might be an example. The solution in these cases is a sliding roof. This can be manually operated using a pole, as in Figure 8.20, or fitted with an electric motor.

High cube vehicles

Some loads are intrinsically dense, such as canned fruit or palletized books, and the vehicle will reach its maximum weight capacity without using the full volume available. However, this is not always the case. Apart from obvious examples such as footballs, some commodities such as car body panels and hanging garments require packaging that incorporates a great deal of free space, hence an artic full load can be surprisingly light, and the vehicle will reach its maximum volume capacity before it reaches its maximum permitted weight.

To overcome this problem, an obvious solution is simply to build a high trailer. However, there are legal limits to this in most EU countries – 4 metres is the maximum

Figure 8.20 Sliding roof on a curtainsided trailer being opened to permit loading from above

SOURCE Reproduced by kind permission of Denby Transport Limited

permitted vehicle height in Belgium, Germany and Italy, for example – and practical limits in the UK due to bridge heights. The additional aerodynamic drag also increases fuel consumption.

Alternatively, the chassis height of the vehicle can be lowered, and the whole of the loading deck reduced in height accordingly: the term low-axle technology is sometimes heard. This has the disadvantage of reducing available payload in terms of weight. For example, the Volvo FH16 is available in a range of chassis heights from 850 to 1200 mm, but maximum front axle loading is just 7.5 tonnes for the former compared with 10 tonnes for the latter (Volvo, 2018).

Some trailers therefore have a step-frame or swan-neck design. Whilst the front part of the trailer is of conventional height, the remainder of the floor is at a lower height, and the trailer wheels are reduced in size accordingly. This option is sometimes combined with a lifting roof, whereby the trailer roof can be raised by perhaps 30 centimetres, and re-lowered after loading, to prevent the loss of space at the very top of the trailer due to the pelmet supporting the curtains. One disadvantage of step-frame trailers, or indeed any reduced size wheels, is that the tyres must make more revolutions per kilometre travelled. This increases tyre wear and therefore tyre replacement costs. It is also important to ensure that the loading deck height is compatible with any loading docks that it will need to use.

Figure 8.21 Step-frame trailer with an additional deck to improve space utilization

SOURCE Reproduced by kind permission of Palletline

Increasing the volume within a trailer will have limited usefulness if the goods are not stackable. This problem can be overcome by using a double-deck (or multiple-deck) trailer. The additional deck may be fixed; foldable to the side; or fitted with hydraulic rams to permit it to be raised or lowered as required. Some companies use trailers incorporating both a step frame and a double deck, either for roll cages or for mixed palletized goods. An example used by Palletline is shown in Figure 8.21.

To increase the capacity of smaller vehicles, a Luton van body is sometimes fitted. This includes a small additional section above the cab, and is sometimes used for local furniture deliveries. In practice, it can be difficult to make use of the additional space due to its small size and awkward position, and it may well be rarely or never used. I would not therefore recommend such a vehicle unless the use to be made of the extra space has been clearly identified.

Skeletal trailers

Seafreight containers are frequently seen on the roads, being hauled by artic units from a port or rail terminal to an unloading point.

The container will be carried on what is called a skeletal trailer, or 'skelly'. The name derives from early versions that comprised a simple steel frame mounted on a chassis, to match the dimensions of a standard 40-foot container, with twist locks at the four corners to secure it to the trailer. More modern and robust versions of such trailers are still produced, and will meet the needs of transporting 40-foot containers but no other sizes.

One problem with such trailers is that if a 20-foot container needs to be moved, it has to be placed at the rear to permit unloading. However, this causes all the weight to be borne by the rear axles, and this may cause legal axle weight limits to be exceeded.

There are two solutions to this problem. One is to use a sliding skeletal trailer. The 20-foot container would be positioned midway along the trailer for transit, to spread the weight more evenly between axles. On arrival, the driver operates a hydraulic mechanism that slides the container to the rear of the trailer for unloading.

An alternative is a skeletal trailer that can be split into two. A 20-foot container can be carried on each of the front and rear halves, and on arrival air operated locking pins can be disconnected so that the trailer splits into two. Each container can then be unloaded separately – at more than one location if necessary (Figure 8.22). The additional flexibility of such trailers is clear to see. One word of warning is that two heavily loaded 20-foot containers will not be able to be transported legally in this way if together they exceed maximum legal GVW.

Figure 8.22 Skeletal trailer. This particular model can be divided so that two 20-foot containers can be unloaded at separate loading docks

SOURCE Reproduced by kind permission of Dennison Trailers Ltd

Tankers

The type of tanker that will be most familiar is the large artic delivering petrol or diesel to a filling station. There are many other commodities also delivered by tanker, including both liquids and powders (Figures 8.23 and 8.24). Examples include:

- lubricating oils;
- liquified gas;
- chemicals;
- milk;
- beer;
- wine;
- chocolate (molten);
- flour;
- salt;
- sugar;
- animal feed;
- sewage and effluent.

Figure 8.23 Tanker used to transport foodstuff grade liquids

SOURCE Reproduced by kind permission of Abbey Logistics

Figure 8.24 Tanker used to transport powders

SOURCE Reproduced by kind permission of Abbey Logistics

Whilst some tankers tend to carry just a single product in a single compartment – such as milk – others may have several compartments: there are many different types of lubricating oil, for example. Even those with a single compartment will usually incorporate baffles to prevent the liquid sloshing around when braking or cornering: this could cause instability, especially when the tanker is only partly full, and be a serious safety hazard.

Unloading may be by pump or gravity, aided by tipping in some cases. Some tankers are metered, so that a specified quantity can be delivered on each drop – domestic heating oil deliveries to a number of different properties would be an example. Some tankers are temperature controlled, either chilled (eg for milk) or heated (eg for molten chocolate).

Tanker cleaning is important, and must include ancillary equipment such as discharging hoses. There are strict rules about certain product groups, and logs must be kept to ensure traceability. Clean-in-place (CIP) is the method most commonly used for internal cleaning – in simple terms, apparatus is lowered into the tank and internal surfaces are sprayed with detergent, then with a rinsing agent.

Refrigerated vehicles (also known as fridges or reefers)

> Refrigerated trucks and trailers carry some of the most valuable and sensitive loads on the road. A reefer laden with smoked salmon could easily be worth many thousands of pounds, while the value of a consignment of pharmaceuticals can reach seven figures.
>
> (Scott Dargan, MD, Transicold, 2018)

With this in mind, it is vital that the integrity of a cool chain is maintained, with goods at a constant optimum temperature throughout. Refrigerated vehicles may be carrying goods that require different temperatures – a supermarket home delivery may include ambient, chilled and frozen goods, for example, and individual pharmaceutical products may each have very specific requirements. Different compartments set to different temperatures are therefore a common feature.

Advance planning is also important. Achieving a desired temperature is not an instantaneous process, and units must be activated in good time to allow conditions to adjust: modern technology will often allow this to be done, and temperatures monitored, remotely. Airtight seals on loading bays to prevent warm air entering the system are highly desirable, and when goods must temporarily leave the cold chain, for example when roll cages are being wheeled from vehicle to freezer at a retail outlet, labour should be available to make the transfer without undue delay.

Refrigerated vehicles require insulation, and an additional width is legally permitted so that this can be accommodated – up to 2.60 metres. The refrigeration units themselves obviously require a power source. This may be an independent diesel engine, or a power take-off from the main engine of the truck, or in some cases both options. Stand trailers may require an external power source – the Port of Dover has 130 such reefer points, for example (Port of Dover, 2017). It is wise to check in advance that these are available.

Either an independent diesel engine or a power take-off will by its nature increase exhaust emissions, and staying within legal limits – for both emissions and night-time noise levels – can be an issue in some cities.

Specialized vehicles

There are a great many specialized vehicle types in use, designed for the transport of one particular type of freight, but with numerous variations within that category. Car transporters, for example, can be designed for a single car or for up to 12. Other specialist uses include the transport of cash, livestock or refuse; delivery of a specific commodity from ice cream to ready-mixed cement; or provision of an emergency service such as fire engines and breakdown vehicles. I will not attempt to describe them all.

Loading and unloading aids

Most vehicles are loaded and unloaded either by hand, or at locations where independent equipment such as a forklift is available. However, if this is not the case, they need to carry suitable equipment.

One option is to carry a forklift. These are usually purpose-designed compact units that can be carried attached to the rear of the vehicle – an example can be seen in Figure 4.1 (page 000). For occasional lifting, a hand-operated lift with a manual winch, of the type illustrated in Figure 4.2 (page 000), can be carried.

Self-propelled items such as cars can be driven onto a vehicle via a ramp. The difficult stage in loading tends to be when moving from the ramp onto the main deck of the vehicle. A car driven onto a transporter too fast may ground against the edge of the deck, causing a dent to the underside. Whilst a car can still be sold as new if certain repairs have been made, such as touching up a small scratch or even replacing a whole door with a new door, such repairs cannot be made to this type of damage and the car will no longer be able to be sold as new. The cost of the incident will therefore be substantial – not all such cars can be used as company cars by the transporter company. Similarly, forklifts tend to have very low ground clearance, and a shallow ramp angle is therefore necessary to load them.

Such loading can be aided by a winch, either manual or mechanical. Examples include broken-down cars being loaded onto recovery vehicles, and boats being hauled out of the water onto trailers.

Tail-lifts

If you need to unload heavy, possibly palletized, items from vehicle deck to ground level, a tail-lift will probably be the best solution. This is a platform that can be raised or lowered by vertical hydraulic rams or connecting arms.

The main types in use are column, cantilever and tuck-away. Colum lifts take their name from columns attached to each side of the rear of the vehicle, from which the rams operate. The platform can only operate at an angle of 90 degrees from the vehicle, which can create difficulties operating on ground that is not level. Cantilever tail-lifts have a hinged platform, which can operate at different angles and therefore overcome this issue but are more expensive.

Tuck-away and slider lifts, as the name implies, can be tucked away or slide beneath the vehicle when not in use (Figure 8.25). They are currently the most popular type as they can be all but forgotten when not required and do not get in the way if the truck is being offloaded via a conventional loading dock.

A range of lift capacities and platform sizes is available – it worth remembering that a heavy-duty tail-lift will have a greater effect on vehicle payload. If you are unloading individual gas bottles then 500 kilogramme capacity will be more than adequate; for heavy palletized goods (perhaps delivering sacks of fertilizer to farms), 1,500 kilogrammes will be more appropriate; and 3,000 kilogrammes will be sufficient for almost any need.

Whilst most tail-lifts operate only to the height of the main vehicle deck, double-deck trailers will require a tail-lift to reach the upper deck. This is often seen on trailers carrying two decks of roll cages. Fences around the edges of the platform are recommended for safety reasons on such vehicles.

It is important to train personnel in the correct use of tail-lifts. Heavy items should be positioned along the centre line of the platform close to the vehicle to avoid damage to the tail-lift: such disciplines are especially important if an item is unevenly weighted, such as a powered pallet truck.

Figure 8.25 Tuck-away tail lift being used to unload pallet

SOURCE Reproduced by kind permission of Palletline

Cranes

The other main option is a vehicle-mounted crane (Figure 8.26). These are used mainly (but not exclusively) in applications where the terrain would make tail-lift operations difficult; and the vehicle is open-topped, such as a flatbed or dropside. Industries include construction, forestry, and some more specialized examples such as portable toilet rental companies.

There is again a huge variety of models on the market. The simplest are manually operated swing lift cranes, which use a lever-operated hydraulic pump. An entry-level model with a lift capacity of 250 kilogrammes will cost just a few hundred pounds. An electric version taking its power from the vehicle battery might be fitted inside a van. Urgent car parts deliveries, usually able to be handled manually but including an occasional engine or gearbox, would be an example of their use. Small swing-lift cranes tend to be light (perhaps only 20–25 kilogrammes) and thus have a minimal effect on payload. Some are designed to be easily demountable and thus able to be fitted or removed at short notice.

At the other end of the market, truck-mounted cranes with a reach in excess of 20 metres or a working load in excess of 30 tonnes are offered, although all models will have reduced lifting capacities at extended boom lengths. Accessories include load sensors, which would seem a good idea. Some cranes simply have a hook to which slings or chains can be attached, but a common lifting attachment is a sideways operating clamp, which can lift, for example, a palletized stack of breeze blocks by clamping either side. Specialized grabs for logs or barrels are available, as are clamshell scoops for materials such as sand and gravel.

Figure 8.26 Crane mounted on a flatbed, delivering to a construction site

SOURCE Reproduced by kind permission of Eddie Stobart Logistics

Vehicles carrying high-capacity cranes must be fitted with outriggers, to be lowered before activating the crane to ensure a stable lifting platform. Thorough training of crane operating personnel is of course essential – failing to fully lower a crane after use, then striking a low bridge, is a more common type of accident than it should be.

Automated systems

It is possible to extend conveyor systems into trucks. Using advanced technology, it is possible for pallets to be loaded directly onto trailers using roller conveyors without use of forklifts or other equipment. Clearly, such systems require extremely large capital investment.

Finally, some trucks are fitted with moving floors, to offload materials such as biomass granules. Again, this is a highly specialized solution.

Conclusion

In this chapter I have described the many types of vehicle available, and I hope this will have enabled you to work out the best solution for the requirements of your specific business.

In future chapters I will discuss the best ways to operate those vehicles.

Abnormal indivisible loads

09

Most freight can be transported on conventional vehicles, as described in the previous chapter, but this is not always the case – occasionally an item is simply too large. This is known as an abnormal indivisible load, or AIL.

UK law recognizes this and makes allowance for their movement subject to special requirements.

> The legal definition of an AIL is 'Loads which cannot, without undue expense or risk of damage, be divided into two or more loads for the purpose of carriage on the road and which cannot be carried on a vehicle operating within the limitations of Construction and Use Regulations' (Lowe and Pidgeon, 2018).

In practice, this means a loaded vehicle that exceeds:

- 44 tonnes gross vehicle weight (GVW); *or*
- axle load of 11,500 (driving axle); *or*
- axle load of 10,000 kilogrammes (non-driving axle); *or*
- 2.9 metres width; *or*
- 18.65 metres length.

(Hallett Silbermann, 2018)

Examples include the rotor for a wind turbine, a seagoing yacht, a ship's boiler, a portable office or a crane for a construction site. All of these may exceed the loaded vehicle expectations above.

It should be noted that the load must be indivisible to be moved as an AIL. To move a single piece of construction machinery weighing 60 tonnes would be legal. To move three such pieces, each of 20 tonnes, might be cheaper on a single vehicle but they must be shipped on three separate conventional vehicles to remain within the law.

Having said this, it is permissible to ship engineering plant and its ancillary equipment on the same vehicle, provided that it is travelling between the same locations and provided the GVW does not exceed 80 tonnes. Examples might include alternative lift attachments removed from a crane, or a bucket removed from an excavator for ease of transport.

AILs under Construction and Use Regulations

Where loads exceed the limits for conventional vehicles, but only by limited amounts, it is possible to operate the vehicle under Construction and Use (C&U) Regulations, ie those laid down for all vehicles operating on public roads. Consider, for example, a radio mast that overhangs the rear of a trailer:

- If it does so by less than 1 metre, there are no special requirements.
- Above 1 metre but below 2 metres, the end must be clearly visible.
- Between 2 metres and 3.05 metres, marker boards must be fitted.
- Above 3.05 metres, an attendant must be carried, and the police must be informed.

(Hallett Silbermann, 2018)

The duty of an attendant is officially to warn the driver and anyone else in the area of any danger that may occur – in practice their main task will be to assist by guiding the driver whilst manoeuvring the vehicle.

A free website, Electronic Service Delivery for Abnormal Loads (ESDAL), is available for notifying the police. The old system of completing forms can still be used but is less efficient and not recommended. ESDAL includes an automated routing system that calculates the best route between two points for a given load, identifies the authorities that must be notified, and does so at the click of a mouse. Other pay-to-use notification services, such as Abhaulier, can be used for the same purpose.

Notice must be given to the police at least two clear working days in advance – ie a load notified on Monday cannot be moved before Thursday.

In addition to the police, for loads exceeding 44 tonnes GVW other authorities including councils and owners of roads and bridges, such as Network Rail, The Canal & River Trust, and Transport for London etc must be notified and an indemnity must be given. If the load exceeds 80 tonnes GVW, these additional authorities must be given five clear working days' notice.

Special Types General Order

For larger loads, a special set of legal regulations apply. These are known as the Special Types General Order (STGO) Regulations. The rules are contained in s.44 of the Road Traffic Act 1988 and the Motor Vehicles (Authorization of Special Types) (General) Order of 2003 (Lowe and Pidgeon, 2018).

There are three categories:

1. STGO Category 1 – up to 50 tonnes. This is rarely used in practice, as few loaded vehicles exceed 44 tonnes but not 50 tonnes.
2. STGO Category 2 – up to 80 tonnes.
3. STGO Category 3 – up to 150 tonnes.

These movements require specialized equipment, which must display a plate showing 'special types use' weights and speed limits. Tractor units often have much more powerful engines than normal – up to 750 HP – and special requirements apply to many other aspects of construction, such as service and parking brakes. The main types used are low loaders, in which the load area is below the level of the platform above the wheels; and semi-low loaders, in which the load area is at the same level. In practice many trailers used for such work can be adjusted, with the load area able to be raised, lowered or extended.

There are usually restrictions as to the times of movement. For example, no movements to the West Country are permitted on summer weekends, and the Metropolitan Police do not permit movements in London between 07.00–10.00 and 16.30–19.00; any movement of over 100 tonnes must take place between 19.00 and 07.00. This can create problems in practice, as planning permission restrictions may prohibit work on site at the times at which equipment can be delivered. This will result in vehicles being delayed for extended periods and in some cases double crews being needed to comply with drivers' hours regulations. Each of these obviously attracts a cost.

Figure 9.1 Low loader vehicle operating under STGO Category 3 for transport of a 70-tonne tracked excavator

SOURCE Reproduced by kind permission of Hallett Silbermann

There will also be additional costs if police or private escorts are needed – whilst there are guidelines, each individual police force can make its own rules as to when escorts are or are not required. There may also be charges if temporary road closures are needed: recently hauliers have been charged £2,800 for a 15-minute local road closure in Hammersmith to permit vehicles to proceed the wrong way in the area's one-way system. Indemnities are also required, to pay for potential damage to roads or bridges caused by an abnormal load.

A further operational problem occurs in the event of traffic incidents, such as motorway closures following accidents. The load is only permitted to follow its authorized route. If a diversion is put in place via other roads, it is illegal for the vehicle to follow that diversion, even if it has been put in place by the police.

> It is vitally important to notify details accurately, as authorizations are very specific. A load notified at 90 tonnes that is actually 92 tonnes will be considered 48 tonnes overweight (ie compared to the 44 tonne limit). In 2017, a fine of £360,000 was imposed in such circumstances. Consignors failing to take into account, for example, of additional equipment such as a winch that has been added to machinery, or worse still under-declaring weights and dimensions in the hope of obtaining a cheaper price, can have extremely serious consequences.

Figure 9.2 Crane loaded aboard a low-loader trailer for transport under STGO Category 3. The tractor unit and trailer have a total of 12 axles

SOURCE Reproduced by kind permission of Hallett Silbermann

Very large consignments, such as those over 150 tonnes GVW, over 30 metres in length, or over 6.1 metres wide require a Special Order. Approval is not guaranteed, and at least 10 weeks' notice should be given.

Finally, there is officially no limit to the height of loads in the UK, or any requirement to notify the movement (unless it is abnormal in other regards), although it would always be advisable to do so. In practice, however, loads must be able to pass safely under bridges, which on motorways are usually a minimum of 4.85 metres but often not more than 4.95 metres.

Vehicle operations 10

Costs and their reduction

It will become clear from reading this chapter that operating goods vehicles is not cheap, and certainly not as cheap as many with no connection to the transport sector tend to believe. It is not unknown to hear customers make statements such as 'You will be delivering on your own vehicle. You already have that vehicle so to deliver goods to us will not cost you anything.' If you are considering operating your own vehicles, be it a single van or a fleet of artics, it is important to understand the cost implications. Even when using a 3PL, it is useful to know whether the costs being quoted are realistic.

In this chapter I will look first at what the overall costs are likely to be, and the major elements to those costs. I will then discuss each of those elements in more detail, giving some advice on how they can be reduced.

Overall costs

Motor Transport, a leading periodical for the road transport industry, publish cost tables annually for different types of vehicle, and four examples are given in Table 10.1.

These are typical figures, but each case will of course be different. Drivers' wages will, for example, differ according to the geographical area. If you are contemplating establishing an in-house transport operation, I would recommend that you perform your own calculations, using these headings but inserting your own figures.

It can be seen that the cost of operating an artic is almost £130,000 per year, or over £500 per working day. The decision to begin your own vehicle operations should not be taken lightly.

I have discussed vehicle specifications, which will largely dictate purchase price, in Chapter 8. There is little more to be said, other than the obvious commercial point that regular large buyers of trucks obtain the cheapest prices. It will always be worthwhile to compare what is on offer from different sources, and if you are only looking for a single vehicle you may be lucky enough to pick up a good deal perhaps due to a cancelled order from elsewhere.

Table 10.1 Vehicle operating costs

	3.5t van		7.5t rigid		18t rigid		44t artic (6x2 + triaxle)	
Vehicle cost	£ 22,666		£ 43,274		£ 64,960		£ 100,738	
Residual value	£ 4,043		£ 7,017		£ 10,485		£ 14,123	
Fixed costs								
Depreciation (years)	£ 3,725	5	£ 7,251	5	£ 10,895	5	£ 11,315	unit 7, trailer 12
Finance cost	£ 653		£ 1,190		£ 1,778		£ 5,893	
Insurance	£ 1,565		£ 1,746		£ 2,216		£ 3,819	
Vehicle tax	£ 140		£ 165		£ 650		£ 1,850	
Driver's wages, NI	£ 24,433		£ 28,876		£ 31,366		£ 35,304	
Establishment/ overheads	£ 5,233		£ 7,299		£ 14,229		£ 23,787	
Profit allowance 5%	£ 1,787		£ 2,326		£ 3,057		£ 4,098	
Total annual fixed cost	**£ 37,536**		**£ 48,854**		**£ 64,191**		**£ 86,066**	

Variable costs								
Fuel (mpg)	15.70	28.0	26.00	17.0	34.00	13.0	52.00	8.5
AdBlue	N/A		0.36		0.48		0.73	
Tyres	1.38		2.24		2.65		5.37	
Maintenance, repairs	4.83		7.19		7.71		11.50	
Profit allowance 5%	1.10		1.79		2.24		3.50	
Total variable costs per mile	**23.00**		**37.50**		**47.00**		**73.10**	
Total annual cost (miles)	**£ 44,436**	30,000	**£ 63,854**	40,000	**£ 87,691**	50,000	**£ 129,926**	60,000

SOURCE Reproduced by kind permission of *Motor Transport* (2018b)

NOTES 1. Depreciation period is that over which the value of the vehicle is assumed to fall from its original cost to the residual value.
2. Depreciation is calculated on a straight line accounting basis, ie in equal amounts each year.
3. Fuel consumption is measured in miles per gallon, eg the van is assumed to use an average of 1 gallon (4.54 litres) for each 28 miles travelled.

I will be discussing insurance later in the book. In this chapter I discuss other important aspects in more detail.

Residual value and depreciation

> When deciding on vehicle options, the effect on residual value should be taken into account. Sleeper cabs, for example, hold their value better than day cabs. However, an upgraded in-cab entertainment system is likely to have little effect on the value of a used truck.

There is of course the option to buy a second-hand vehicle. However, this is fraught with danger for the inexperienced buyer – it can be very difficult to spot whether a truck has been well maintained throughout its working life, or whether the bare minimum has been done to meet legal requirements and the vehicle cosmetically smartened up for sale.

Recognizing a bargain is as much an art as a science. If you have the expertise to do so within your organization, then by all means take advantage of this. If not, at the very least I would recommend paying for an independent inspection before committing to buy. Better still, try to structure your purchase to include an extended guarantee and service contract to protect against the risk of major unplanned expenditure.

In Table 10.1, depreciation is assumed to be on a straight-line basis, and in deciding the viability of a planned purchase I would advise that this method is used so that the project can be costed over the planned life of the vehicle. However, this is not a reflection of actual vehicle values. Company accounting policies may dictate use of the 'sum of the digits' methodology. This weights depreciation towards the earlier part of the vehicle life and if there is the possibility of the vehicle needing to be sold earlier than intended, this will give a better reflection of value at any given time. Other accounting conventions exist, and you should consult with your company's accountant or finance director at an early stage of the process.

Finance

About three-quarters of UK truck purchases involve some type of credit funding. Of the remainder, most are purchased by truck rental companies and the like, so few are outright cash purchases by the user company (Commercial Motor, 2018). The main reason is that few firms are sufficiently cash rich to spend a million pounds on a fleet of ten artics, and those that are may well find they can obtain a better return on their capital elsewhere. Purchasing the freehold of your transport yard is likely to be a better long-term investment than purchasing the vehicles parked in it.

Also, if you are buying just a small number of vehicles, maybe only one, your purchasing power with the manufacturer will not be that great. It may be that buying through a major vehicle rental company and taking advantage of their purchasing power will outweigh the profit margin of the middle man.

Against this, small businesses should be aware of the tax benefits available from the Annual Investment Allowance and Writing Down Allowance. The former allows the first £25,000 capital investment in a tax year to be offset against tax; the latter allows 18 per cent of capital investment not covered by the Annual Investment Allowance to be offset against tax each year (Commercial Motor, 2018). If you are a small business who wants to set up a small fleet, maybe only a single van, these allowances may make outright purchase much more attractive.

Spot hire

At the other extreme, vehicles can be spot hired on a monthly, weekly or even daily basis, and even major operators are likely to do so from time to time. It should be remembered that even for a single day's hire of a vehicle over 3.5 tonnes, an operator's licence will still be required. The main advantage is that there is no commitment, and the vehicle can be returned at the end of the hire period without penalty (unless it is damaged). Mechanical repairs and servicing remain the responsibility of the hire company. Payment can usually be made upfront, which avoids the need for creditworthiness checks, which can be useful to a new business. The disadvantage is of course cost – this is the most expensive way to fund vehicles. Also, it can be difficult to obtain vehicles at busy times – anyone who has tried to spot hire artic trailers in the weeks before Christmas will tell you this can be far from easy.

When spot hiring vehicles, be sure to inspect carefully on collection and report damage found. Similarly, ensure that the vehicle is inspected on return before handover. Some unscrupulous hire companies attempt to charge a succession of customers for 'repairing' the same item of damage.

Longer-term agreements: General comments

Creditworthiness checks will be required if you want to sign a longer-term agreement. The lease or hire company will quite reasonably want to be sure that you can meet payments over the whole period of the agreement, and in some cases guarantees will be required. This can be a problem for a new business.

From your viewpoint, it is also not unreasonable to seek protection from being supplied with a dud vehicle. When dealing with a major manufacturer who supplies a lengthy warranty, this is unlikely to be a problem. However, there have been cases where a small independent dealer has provided a second-hand vehicle, perhaps even with a guarantee, then ceased trading, leaving the guarantee worthless. Finance is provided by an independent finance institution, which will insist on full payment even if the vehicle has so many defects that it cannot be economically repaired.

A finance agreement that includes phrases such as 'As we have no role in selecting the vehicle we can take no responsibility for its suitability' is a warning sign. If you do not have protection from elsewhere, you may wish to walk away.

Replacements whilst a vehicle is off the road due to breakdown or accident are usually an optional extra. The contract may state that they will be provided free at the point of use (though obviously the base price reflects this) or spot hire rates may be charged as appropriate. Leasing companies usually set the price so that the overall cost is very similar whichever option is chosen, so the choice becomes largely a matter of personal preference.

The agreement will most likely include a maximum mileage, perhaps 50,000 per year over three years. Any additional mileage will be charged at a pre-agreed figure of X pence per mile. If possible, when hiring several vehicles together agree a pooled mileage figure, so that excess mileage on one vehicle can be offset against lower mileage on another – this avoids the need to swop vehicles around between routes to avoid charges.

The agreement will probably state that at the end of the period the vehicle should be returned free of damage, but fair wear and tear is acceptable. The British Vehicle Rental and Leasing Association defines this as:

> Fair wear and tear occurs when normal usage causes deterioration to a vehicle. It is not to be confused with damage which occurs as a result of a specific event or series of events such as impact, inappropriate stowing of items, harsh treatment, negligent acts or omissions.
>
> (BVRLA, 2018)

The BVRLA also issues a more detailed standard. You should obtain a copy from the leasing company, or you can purchase a copy from the Association's website. They will provide an arbitration service if necessary, but there is obviously a cost associated with this. It is wise to ensure that there are no missing items such as spare keys or service history, and it may be worthwhile to repair minor damage in advance. If specific items are added at the beginning of or during the contract, such as livery or accessories, it should be agreed at the time whether or not charges will be made for their removal.

Finally, clean the vehicle inside and out immediately prior to return. Not only will this help to get the vehicle inspector in a sympathetic mood but it will avoid what can be very high charges for a professional valet.

Contract hire

The main difference between spot and contract hire is that there is a minimum commitment – usually between one and five years (some people use 364-day hire periods, as a year or above would require board approval; I would strongly discourage such procedure avoidance). Mechanical repairs remain the responsibility of the hire company, which is attractive to organizations that wish to avoid risk. The longer-term commitment will yield substantial savings against spot hire, but penalties for early termination of the agreement are likely to be substantial.

Operating lease

As with contract hire, there will be a minimum term, again usually between one and five years. The difference is that you will take responsibility for servicing and mechanical repairs – although you can make an agreement with a separate organization to provide this. At the end of the contract period the vehicle is returned to the leasing company, and its future disposal, including the risk that it may not reach its expected residual value, is their responsibility.

Hire purchase and alternatives

Most people will be familiar with hire purchase (HP). A deposit is paid, perhaps of 10 per cent, then regular payments over a number of years to cover the whole cost of the truck, plus interest. You will be able to claim capital allowances against tax. At the end of the contract a small admin fee is paid, and the truck becomes yours. A finance lease is similar, except that you do not gain legal ownership. At the end of the period you can sell the vehicle and keep most of the proceeds – typically 95 per cent – or extend the agreement into a secondary lease with just a token monthly payment. As you will never have ownership it is the leasing company that is able to claim capital allowances against tax, and this should result in lower monthly charges than HP (Commercial Motor, 2018). Under a contract purchase agreement, monthly payments are lower than for HP, but a final payment equivalent to the anticipated residual value of the vehicle must be paid to secure ownership. Returning the vehicle to the lease company at this point is an option, so the decision can be made at the time based on whether the market value is higher or lower than expected.

All of these methods must be reflected in the company's balance sheet, and both assets and liabilities will need to be shown. If you wish to keep vehicles off the balance sheet, for example because you believe it will make it easier to obtain further credit, then an operating lease is a better choice.

Road tax

The rules for UK Vehicle Excise Duty, otherwise known as road tax, have become increasingly complex over recent years. The basic aim is to set rates that reflect the amount of damage done to roads and the environment, such that the most damaging vehicles pay the highest amounts. Charges are higher for vehicles with higher permitted weights but may be reduced if there are more axles and/or road friendly suspension, and/or emissions are lower. For vehicles over 12 tonnes there is a levy to be paid, which is also paid by foreign registered vehicles entering the UK. Overall, this leads to a highly complex system. Lowe and Pidgeon (2018) provide comprehensive tables extending over 15 pages of their book, and my best advice is to determine the tax level for the particular vehicles that you are considering, and to take this into account in your decision.

If you are sure you will not be operating at the maximum weight for your vehicle, it is possible to voluntarily down-plate a vehicle, ie reduce its maximum permitted weight. This can be done at a local goods vehicle testing station and is administered by sending a Notifiable Alterations Form (VTG10) to the DVLA at Swansea. The fee will be £27 or £40. However, down-plating an 18 tonne rigid to 12 tonne will reduce excise duty from £650 to £200 per year.

Drivers' pay

This is one area where reluctance to pay a reasonable rate is likely to be counterproductive. There has been a shortage of drivers in the UK for many years. A Department for Transport enquiry in 2015 concluded that there were about 45,000 vacancies but fewer than 1,200 drivers claiming Job Seeker's Allowance.

The average age of drivers was 48, and the problem is therefore likely to worsen as an ageing workforce retires but insufficient young people join the industry to replace them (UK Parliament, 2015). The reasons for this are many, but may include poor working conditions, a negative public image, and/or a not unnatural desire to spend time at home with one's partner and children rather than sleeping in a cab.

If you are to attract a good standard of driver you will need to pay a competitive wage. This will be especially true: if you need to impose antisocial shift patterns to meet the needs of your customers (for example, 3am starts to meet 6am deliveries); if you are in an area where there is great demand for both HGV and coach drivers, such as around Heathrow; and/or if specialist skills are needed. Car transporter and tanker drivers have historically attracted higher wages for the latter reason.

When costing this element, National Insurance and pension contributions must be included, along with cover for holidays and sickness. The former can to some extent be managed, for example by stipulating that part of the annual entitlement must be taken at times of low demand, such as factory shutdowns or between Christmas and New Year, by forbidding holidays at particularly busy times, and/or by paying bonuses only if holidays are taken as whole weeks rather than individual days. Managing sickness absence is more difficult, but no less important, and high staff turnover creating frequent vacancies is most undesirable.

For a large fleet, it will probably be viable to employ spare drivers to act as a reserve and step in wherever necessary. However, few transport operations are able to avoid the use of temporary drivers from an agency at some point. Some of these are agency drivers by choice – they may enjoy the variety of different work or be semi-retired and not need to work every week. However, others are inexperienced drivers or even those unable to hold down a permanent job for whatever reason. At busy times, only the latter may be available, and at very busy times it may be impossible to find any agency drivers at short notice. Temporary drivers will be less familiar with your particular needs and cannot be expected to have a passion for your brand. It would certainly be unfair to taint every agency driver, and I do not wish to cause undue offence, but my own experience is that they perform less well, and they are anecdotally more likely to cause vehicle damage, make late deliveries, or create a poor impression at customer premises. I would therefore recommend building a strong relationship with a particular agency and liaise with them closely as to which individual agency drivers perform best. Minimizing use of agency labour by proper management of your own drivers, maintaining the right balance by treating them well whilst not tolerating laxity are essential to a cost-effective transport operation.

Establishment, overheads and profit allowance

One of the most difficult costs to quantify is – put simply – which costs do you count?

For a small operation, just a single van in some cases, the incremental costs will be minimal. A van can be obtained, and a driver hired, and they can be managed without any additional resources. Someone such as the warehouse supervisor can be responsible for line management along with their other duties, and the personnel department will barely notice the effect of a single additional employee.

There will, however, come a point at which additional resources are required. Maybe a transport supervisor will need to be employed, with the skills and experience to manage the fleet and drivers to best effect. A crew room and/or transport office may be needed, accounts will need to process more invoices and additional wage payments, senior management time may need to be expended, and additional phone and electricity usage may be extremely difficult to quantify. Costs such as operator's licence fees, drivers' uniforms, and tachograph analysis may be unavoidable.

In such circumstances my best advice is to take a realistic view of each element. If managing the drivers takes a quarter of an administrator's time, then a quarter of the cost of that administrator needs to be costed in, and so forth. Sundry items should be accounted separately and costed as such in future budgets. There is no right answer in this category – you must simply come to the best estimate you can.

Finally, it will be noted that a 5 per cent profit element is included in the Motor Transport figures. Every business must make a profit to survive, and the transport operation should make at least some contribution towards this.

Fuel purchase

> Fuel is one of the biggest costs in any transport operation – for a 44-tonne artic in the UK this can equate to 52 pence per mile (Motor Transport, 2018b), or an annual cost of over £40,000 per year based on 80,000 miles.

A saving in purchase cost of just a few pence per litre can therefore make a big difference over the course of a year. Even if you are running a low number of small vans, old fashioned methods of making fuel purchases at the pump and claiming back the cost on expenses will encourage purchase at high-cost sites near to drivers' homes, and/or which pay the most generous loyalty points. The cost of administration is also considerable.

I would therefore recommend a system of bulk purchase and/or fuel cards.

Fuel card schemes are offered by most of the major fuel suppliers, by some manufacturers such as Scania, and by independent organizations. For a small fleet, I would suggest combining fuel purchases for company cars with that for vans and trucks to obtain an economy of scale – for very small volumes some card suppliers actually charge more than the pump price for the privilege of using their card. Also, whilst a network of several thousand sites may sound impressive, there may well be gaps in that network, and if a gap happens to be close to your main site, or the only sites close by have low canopies and will not accommodate your vehicles, then this will be a problem. Many companies offer a choice of UK only or Europe-wide card usage.

Security is also a consideration. All cards should be PIN protected, and I would recommend restricting each card to a specific vehicle. Using a company fuel card to fill up a partner's private car is not unknown! Alerts to suspected misuse are offered by some card suppliers, and are a good idea.

In short, I would recommend shopping around, and selecting the best card for your particular business.

For larger fleets, it is wise to buy in bulk, which will yield significant savings. This might be from a major oil company or from a specialist wholesaler. Deliveries can be made into tanks at your operating centres, or into a bunker network such as UK Fuels. The former will of course require substantial capital investment if not already in place, and a strict usage monitoring system is recommended to prevent abuse. Metered deliveries can be made in volumes of 10,000 to 37,000 litres, but full loads are preferable from the cost viewpoint.

The bunker networks operate rather like a bank. The supplier might deliver a full load to a site in Birmingham, then sell this to you. One of your vehicles using a fuel card from the bunker network can then draw 250 litres from a site in Glasgow the following day, and when the stock has been used up by drawing from various sites across the country, another full load will be purchased.

Various pricing mechanisms are available for bulk purchases. My own preference is to base pricing on an independent index such as the Rotterdam Platts, which monitors fluctuations in the wholesale price. An agreed formula might be that the price to be charged for fuel during a calendar week will be based on the midpoint Platts as at noon on the previous Friday, converted from dollars to pounds at the rate published in the *Financial Times*, plus UK Fuel Duty, plus X pence per litre. Such a mechanism gives transparency to all parties. Others may prefer spot pricing, or for prices to be fixed for a period in advance, or for changes to be triggered only when the index increases or decreases by a pre-defined amount. Again, it is a question of selecting the best mechanism for your particular business.

Fuel economy

Having purchased fuel at the cheapest price per litre, the next challenge is to use as few litres as possible. Common sense dictates that this is good practice from both profitability and environmental viewpoints. It is not an exact science as there are a great many variables – even the difference between winter and summer weather can cause a 10 per cent variation in fuel usage (Department for Transport, 2010). Having said this, steps can be taken to significantly improve fuel economy.

The first step is to specify vehicles correctly. This was discussed at length in Chapter 8, but it is worth reiterating the simple point that big trucks use more fuel than small trucks – selecting a larger vehicle than necessary will therefore incur unnecessary expense. It is also worth remembering that a high-roof trailer will increase consumption, but fitting air management systems and lifting axles will reduce consumption. Also, tyres are important: low-rolling-resistance tyres can improve economy by up to 6 per cent on long-haul work (Department for Transport, 2010), and unnecessarily wide tyres should be avoided.

Second, vehicles should be properly adjusted and maintained. Failing to keep tyres inflated to the recommended pressure will increase consumption, but over-inflation, whilst it might slightly reduce fuel usage, will increase tyre wear and most likely prove a false economy. Roof-mounted air deflectors may also require adjustment according to trailer height – a tide mark on the front of the trailer may be a sign that the deflector is set too low. Similarly, one of the benefits of a sliding fifth wheel is that it can be adjusted so that the gap between tractor unit and trailer is kept as small as possible – although it is important to bear axle weights in mind.

Body damage and tears in trailer curtains will increase aerodynamic drag, and they should therefore be repaired as soon as possible. Running with open curtains will quite dramatically do so (as well as exposing the interior of the trailer to dirt and rain). Indeed, properly sheeting an empty tipper body can improve fuel economy by 8 per cent at 56 miles per hour (mph) (Department for Transport, 2010). Curtains should be properly tensioned; straps and buckles should not be allowed to flap in the wind; air horns, roof lights, flag, and other fitments that disrupt clean airflow should be avoided; and roof racks should be removed when not required.

Proper servicing and maintenance are obvious recommendations. Binding brakes, misaligned wheels, dirty fuel filters, leaking pipes and clouds of black smoke are some possible consequences of skimping on proper care of vehicles – and may be a breach of operator's licence requirements as well as a false economy. As part of the process, drivers should conduct a daily routine check before they begin driving, and any defects found should be reported promptly – and rectified promptly, lest drivers reach the conclusion that there is no point in doing so.

Perhaps most importantly, driving style has a very major impact on fuel economy. Many organizations have found it beneficial to send their drivers on fuel economy courses. Often, experienced drivers do not like the idea, and may even feel insulted by the suggestion, but there can be few who do not benefit in some way from such tuition. Monitoring individual drivers' overall performance is also recommended. Whilst all cannot be expected to achieve the same average mileage per gallon, as there are likely to be differences in work pattern (motorway versus urban driving for example), large unexpected differences can be revealed, and addressed accordingly.

Bad driving habits include revving the engine excessively, especially when stationary; use of wrong gears; rapid acceleration; and failing to anticipate the need to stop, resulting in rapid braking. Good habits include always being in the right gear, especially at low speed; keeping an eye on the tachometer (rev counter) dial and making sure the revs remain in the green zone where the engine is most efficient; using cruise control; stopping the engine when stationary; and using exhaust braking and other systems correctly.

Finally, fuel can be saved by the simple expedient of driving more slowly, and some firms adjust speed limiters on their vehicles (and/or fit then when not a legal requirement) to enforce this. Lowe and Pidgeon (2018) quote a study that found that an artic travelling at 40 mph on the motorway will use about 20 per cent less fuel than one driving at 50 mph: 10.5 miles per gallon (mpg) rather than 7.9 mpg. However, it will also cover 20 per cent fewer miles, with a corresponding increase in fixed cost per mile – vehicle depreciation, driver's pay, etc. Based on the figures for a 44 tonne 6x2 artic with triaxle trailer in the Motor Transport tables above, and assuming 45 hours per week of driving 50 weeks per year, overall costs per mile can be calculated as in Table 10.2.

Table 10.2 Cost per mile at 40mph and 50mph

			40 MPH		50 MPH
Fixed costs per year		£ 86,066		£ 86,066	
Per hour		£ 38.25		£ 38.25	
Per mile			£ 0.96		£ 0.77
Fuel per gallon	£ 4.41				
mpg		10.5	£ 0.42	7.9	£ 0.56
AdBlue per gallon	£ 1.55				
mpg (4% of fuel)		262.5	£ 0.01	197.5	£ 0.01
Maintenance, tyres			£ 0.17		£ 0.17
Overall cost per mile			**£ 1.55**		**£ 1.50**

It can be seen that in this example, even with the optimistic assumption that the vehicle will be driven 45 hours per week, the cost penalty of lost productivity by driving more slowly more than outweighs the benefits of fuel economy. There may also be commercial implications of late deliveries if a driver is delayed by traffic congestion and unable to make up for lost time. From the purely financial viewpoint this is not an option I would personally favour; I would recommend good driver training as a better method of improving fuel economy.

Tyres

For the average car driver, a new set of tyres is needed once in a blue moon, and apart from being a nuisance and an unwelcome cost it really receives little additional attention. However, for commercial vehicles the costs are significant. For an artic and trailer, tyre wear can equate to 5.37 pence per mile, and for a 32 tonne tipper 7.74 pence per mile (Motor Transport, 2018b). If double shifting vehicles, this amounts to over £5,000 per year.

To minimize tyre wear, the most important step is to maintain the correct tyre pressure. Over-inflation will cause excessive wear around the central section; under-inflation will cause excessive wear around the outer sections. Excessive wear can also be caused by poor vehicle maintenance: misalignment of wheels, poorly balanced wheels and faulty suspension will all take their toll. Finally, tyre wear is also increased by erratic and aggressive driving. Rapid acceleration, violent braking and cornering at high speeds should all be avoided – not only will they increase tyre costs but also fuel costs, and in many cases the cost of accident damage.

With regard to purchase of tyres, I would recommend sticking with reputable brands. It is possible to buy budget tyres much more cheaply than leading brands, but often these are not manufactured to the same standard and will wear out faster (ATS, 2018). They may be adequate for a yard shunter, or perhaps even for a van that never leaves town and only covers a few thousand miles per year at urban speeds. However, for serious commercial use including extensive motorway driving, a good quality tyre is essential.

Finally, it is wise to ensure that you have a contingency plan in place for emergency call-outs to the roadside. This will normally entail setting up an account with one of the major tyre-fitting companies so that there can be an immediate response if necessary. If a vehicle suffers a puncture on the motorway outside normal working hours, the wheel needs to be changed quickly to avoid a late customer delivery. It has been known for this question to be ignored until the emergency occurs, resulting in extensive desperate phone calls to arrange payment by a personal credit card, this being the only method available. Panic management such as this is most definitely to be avoided.

Maintenance

In times past, maintenance of trucks was a fundamentally simple affair. An experienced mechanic could identify a problem and fit replacement parts as necessary using basic tools and an inspection pit, and many small hauliers had their own workshop. However, this is not now the case. Trucks (and indeed cars and vans) have sophisticated engine management systems, and fault identification is by specialized diagnostic equipment. It is still possible to maintain your own vehicles but establishing a workshop will require substantial investment both in equipment and in a workforce with the necessary level of expertise. I would not therefore recommend this option to an inexperienced operator.

A new vehicle will have a manufacturer's guarantee, and it is usually possible to extend this at a reasonable cost. It is important to ensure that it permits sufficient mileage to cover commercial use, and to read the small print: an unexpected exclusion might appear two-thirds of the way down page 9 of 15 pages of conditions.

I would recommend that a service contract be purchased at the same time as the vehicle, which will cover routine servicing and any mechanical repairs not covered by the guarantee. All other factors being equal, I would recommend sourcing vehicle, guarantee and maintenance together: this will happen automatically in a contract hire agreement. This avoids the potential of disputes as to whether a particular repair should be covered by a guarantee, though this is less likely to occur if the service contract is with a reputable provider.

Before finalizing the agreement, you should establish where and when maintenance will be carried out, and whether vehicles will be collected from your base or whether you will need to arrange their delivery and return. This may be an important factor in your sourcing decision.

Accident damage repairs will not be covered by your service contract. The avoidance of such damage will be discussed in the next section, but it is important to put in place a procedure to control the cost of their repair. I recommend that all repairs should be pre-approved, including their cost, and alternative quotations sought to benchmark the level of charges. It will often be cheaper to arrange for small dents to be repaired by a visiting mobile repair service than by a vehicle manufacturer at their dealership. If not properly monitored, it can be very easy for charges to slip through unchallenged, and soon become a significant problem.

Daily walk around checks

An important part of a driver's job is to complete a walk around check of the vehicle at the start of each shift. It should include:

- tyres:
  - correct pressure;
  - tread depth;
  - no visible worn patches, bulges or damage;
- wheel nut indicators in line (see below);
- load, doors and curtains properly secured;
- mirrors properly aligned, clean, free from frost, and undamaged;
- lights all working;
- wipers and washers working;
- fluid levels sufficient for shift:
  - oil;
  - water;
  - windscreen washers;
  - brake fluid;
- brakes and steering working correctly;
- visible damage;
- any other defect found.

There should be a clear procedure for reporting defects, either found during the check or which become apparent during the working day, and for acting upon them. Drivers who have reported defects in the past only for them to remain unrepaired will be less diligent about doing so in the future.

Cleaning

A dirty vehicle will not be good for your company's image, especially if it has attracted crude graffiti, and both funding and time for cleaning should be part of operational planning.

Commercial facilities, of either the hand-held pressure washer or drive-through type are one possible solution. Alternatively, it is possible to install your own wash facility, a pressure wash system perhaps being most appropriate for a small operation. This will probably mean that vehicles are kept cleaner than if a trip to a commercial facility is required.

Figure 10.1 Static cabinet-mounted pressure washer

SOURCE Reproduced by kind permission of Brendeck Limited

An entry-level system might be static (Figure 10.1) or mobile. The former will have its own water tank, which will refill from the mains, controlled by a ball valve. Power is from an electric motor, and very high pressure (up to 350 bar, for the technically minded) can be generated on some models. On hot models, water may be heated by diesel from an onboard or external tank, or there is an electrical option for indoor use. Mobile units are mounted on a wheeled trolley. An entry-level model will cost a little over £1,000 for a mobile unit, and a little under £2,000 for a static model.

Please note that it may be necessary to install interceptors or separators to ensure that grease, oil and other pollutants do not enter the drainage system. You should consult your sewage utility provider before making any installation.

Wheel nut indicators

It is often said that the best ideas are simple. An example is a wheel nut indicator, which costs just a few pence (Figure 10.2). This is a plastic device comprising a ring and a point. One indicator is fitted over each nut on a wheel and the points aligned. If a nut starts to work loose, the point will come out of line and it will be clear that the nut needs to be tightened.

Figure 10.2 Wheel nut indicators around two wheels of a vehicle

SOURCE Reproduced by kind permission of Checkpoint Safety

Driving standards and accident prevention

Safe and careful driving should be the top priority for any transport operation. There is an unarguable case from both ethical and commercial viewpoints. Road accidents account for 39 per cent of workplace deaths in the EU, and the annual cost of at-work road crashes in the UK is estimated at £2.7 billion: it is estimated that 95 per cent of accidents are due to driver error (Brake, 2018a). This is in addition to the costs I have already mentioned, for example the increases in fuel and tyre costs due to aggressive driving.

Steps that can be taken include:

- Instil a road safety culture throughout the organization. This applies equally to the sales director visiting a potential customer and to a van or artic driver. Posters and intranet messages can help, but word of mouth is the most important communication method.
- Prepare a drivers' handbook and update it regularly. It should include all relevant information from legal requirements to instructions for serving particular customers.

- When recruiting, assess attitudes to safe driving, either through structured interviews or psychometric testing.
- Take up references.
- Obtain a health declaration from all drivers.
- Implement a policy under which drivers are encouraged to report and seek treatment for health problems that affect driving (such as sleep apnoea), and support them, if necessary, with alternative duties whilst they do. A policy that results in immediate dismissal will most likely result in their continued driving despite the dangers.
- Insist on eye tests every two years.
- Establish a policy for the maximum number of points permitted on a driving licence and enforce this (see below).
- Establish a drink and drugs policy (see below).
- Give training to new drivers as part of their induction.
- Establish policies regarding in-cab distractions. Examples include no use of mobile phones, even if hands off, and no unauthorized passengers.
- Install telematics in vehicles, to monitor driving standards. Ensure feedback is given and recorded where necessary, and consider disciplinary action if no improvement is seen. Telematics can reduce safety-related incidents by up to 50 per cent, and many organizations recoup the cost of the systems within a year (Brake, 2018a).
- Maintain vehicles correctly.
- Do not route vehicles such that drivers need to drive aggressively in order to meet deadlines.

The primary aim of the charity Brake is to improve road safety, and they offer useful support material on their website.

Driving licences

Many companies stipulate a maximum number of points on driving licences for their drivers, and it is also a requirement of some insurance policies. In the past, this could be checked by examining the licence itself, but this is no longer possible.

Points can be checked by going to the website www.gov.uk/view-driving-licence. Details of the driving licence, including points and qualifications, can be found online by inputting licence number, National Insurance number and postcode. You can also obtain an eight-character code that will enable someone else to view your details – this is the method I would recommend for potential employers.

Alcohol and drugs

One in six UK road deaths are caused by drivers over the legal alcohol limit, and even below this limit driving ability can be seriously impaired (Brake, 2018b). I would recommend a zero-tolerance policy with regard to drink-driving, with an in-company limit set below the legal limit in England and Wales of 80 milligrams per millilitre of blood. It could be set at 50 milligrams per millilitre – the legal limit in Scotland and many European Countries – or 20 milligrams per millilitre, the legal limit in Sweden and effectively a zero limit for consumption. It should be noted that some countries such as Switzerland and Austria have lower legal limits for commercial drivers than for car drivers. An absolute zero limit is not feasible, as alcohol can be produced by some medical conditions (including side-effects of diabetes) and small amounts may be accidentally ingested as a constituent of an over-the-counter medicine or even in a slightly fermented apple.

Similarly, it would be grossly irresponsible to allow anyone to drive whilst under the influence of illegal drugs, and again I would recommend a zero-tolerance policy. There will be benefits not only from accident reduction but also from apportionment of responsibility. I well recall one forklift driver who was injured and told everyone proudly that he intended to sue the company. He did not pursue this idea after testing positive for both cannabis and heroin.

To enforce these rules, it is possible to buy workplace drug and alcohol testing kits (Figure 10.3). For alcohol, typically this would include a single-use breathalyser that would be used as the initial screen. If this tests positive, an accurate breath alcohol reading can be obtained using a reader of a type similar to those used by the police at the roadside. For drugs, urine or saliva tests can be used for the initial screen. If this proves positive for one of a range of substances, the employee should be suspended on full pay and the sample sent for laboratory analysis to confirm the result.

It should be noted that employees have the right to refuse testing unless they have given consent. I would recommend consent as part of the contract of employment – a potential employee who refuses this would certainly raise suspicions.

The policy should apply to all company car drivers, including sales reps and even directors. I would also include forklift drivers and anyone operating machinery, and there is a case for including office staff. Testing should certainly take place after each accident, and when there is a reasonable suspicion, such as smelling alcohol on a driver's breath. There may even be circumstances in which everyone on a site is tested, such as the finding of a used syringe. I also recommend random testing, either of individuals or of all present at a particular time. To ensure fairness, times, dates and the individuals to be tested should be system-generated.

There may be drivers who develop a problem with drug addiction or alcoholism. These are serious and potentially fatal illnesses. Anyone suffering from them should be encouraged to admit this and seek treatment immediately, whether by counselling, cognitive behaviour therapy, the 12 step programs of Alcoholics and Narcotics

Figure 10.3 Drug and alcohol testing kit designed for the workplace

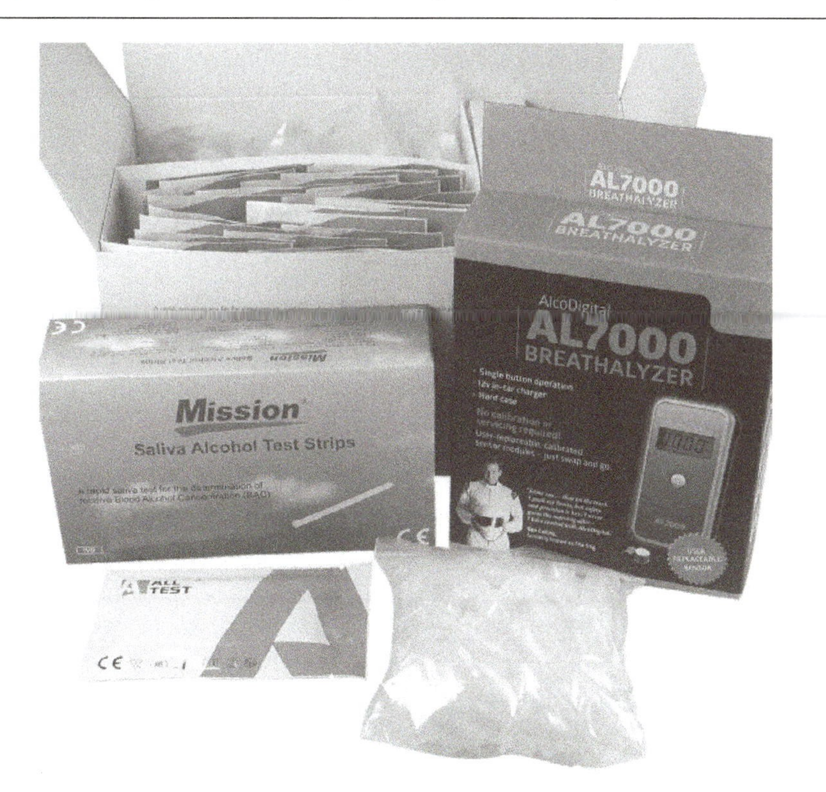

SOURCE Reproduced by kind permission of www.ukdrugtesting.co.uk

Anonymous, or even residential care. As with any other illness, a policy that results in immediate dismissal will most likely result in continued drink or drug driving, which could have tragic consequences. Support during treatment, if necessary, backed by daily testing to enforce a zero-tolerance policy, is a much better approach.

Other costs

Livery and driver uniforms

A smart livery on your vehicles will enhance your company's image, especially whilst delivering to your customers. Brand awareness advertising as the vehicle travels the roads will be another advantage. There will obviously be a cost, which will almost certainly run into four figures for each vehicle. If you are using contract hire, the cost should be agreed as part of the package, including removal at the end of the contract period. A similar principle should apply if you are using a 3PL, although some contracts include a clause to say that any reasonable change to livery will be made by the 3PL on request, at the customer's expense, for example if a new marketing slogan is introduced. However, a livery on a vehicle that is dirty and/or in poor condition will create a negative image, and should obviously be avoided.

In some cases, a livery is not recommended. Examples include wines and spirits, which might make the vehicle a target for robbery, and the defence sector. Also, it is best to avoid political slogans. One retailer once carried messages on its trucks attacking a foreign government over the mysterious disappearance of an opposition leader. Whatever the rights and wrongs of the events, I would fear for the safety of the driver of the vehicle.

Similarly, a smart uniform will enhance a company's image, and smart workwear will also do so. The uniform should include safety footwear, thick gloves, and warm and waterproof outer garments if the driver will be working outside. Again, a scruffy uniform, or an unshaven driver with facial piercings, may not create the correct image, but dismissing a driver for removing a tie whilst changing a wheel is, in my opinion, taking matters too far.

Tolls and congestion charges

If you are operating a small fleet of vans in and around a provincial town, these costs will not arise. However, for long-distance work they can be difficult to avoid.

In the UK there are tolls at a small number of bridges and tunnels, and also on the M6 Toll Road, which effectively acts as a bypass to the often-congested main M6 route through the Birmingham area. In some cases, eg the Tyne Tunnel, it is still possible to pay by cash at a toll booth, but this is not possible at the Dartford Crossing or the Mersey Gateway Bridge opened in 2018, where one-off payments must be made online or by telephone. Enforcement is by number plate recognition cameras.

However, in almost every case it is possible to set up a prepayment account, and I would recommend doing so. At Dartford, for example, you can set up an account, and register vehicles authorized to use that account. It is debited every time a vehicle uses the crossing and can be set to top up automatically when required. There is a direct cost saving – an artic will be charged £5.19 rather than £6 per crossing – and savings in administration. Fines when people forget to make a payment should be avoided, although they can still be incurred if an administrator fails to register a new vehicle.

A similar principle applies in France, where tolls are charged on most motorways. Payment can be made by cash or card at a toll booth, but there are also automated tag-based systems operated by Liber-t and Tis-pl for frequent users. In Belgium, tolls are charged only on vehicles over 3.5 tonnes (apart from a tunnel in Antwerp) for which chargeable vehicles must have an electrotonic on-board device. In Germany tolls are charged on vehicles over 7.5 tonnes on motorways and main roads, payable by on-board device or mobile phone app. However, in Bulgaria you must obtain vignettes at the border. There are plans for a system to cover most countries in Europe, but as of 2018 these have yet to come to fruition. The overall message must be to research carefully before operating in each country.

Congestion charges are imposed in some inner-city areas, including London, Durham, Stockholm, Riga and Valetta. These may be charges at a flat rate, but in some cases are related to emission levels to encourage use of less polluting vehicles. In London, for example, an Ultra-Low Emission Zone will be introduced and extended to the North and South Circular Roads by 2021 (TfL, 2018a). Trucks failing to reach the standard will be charged £100 per day, which will make their use within the area commercially non-viable.

Conclusion

As I said at the beginning of the chapter, transport operations are not cheap. I have laid out the costs that are necessary, suggested how you can quantify them for your particular operation, and provided some ideas for their reduction.

The main cost areas to consider include:

- selection of the correct vehicle – no larger than that required;
- the best method of financing the vehicle;
- road tax, which can be reduced by using a road-friendly and environmentally friendly vehicle;
- fuel, both its sourcing and its economic use;
- maintenance and accident prevention;
- sundry costs such as tolls and congestion charges.

This information will also be useful for benchmarking a 3PL and ensuring the prices they are charging are realistic.

In the next chapter I will discuss the legal requirements for a transport operation.

Legal requirements and compliance

11

The law relating to road transport operations can be very complex, with regulations governing just about every aspect you can think of. There tend to be exceptions to rules too, which makes matters even more difficult at times.

However, some of the rules do not really create a problem in practice. There are construction and use laws for all vehicles, for example, governing quite small details such as the minimum height above ground for an indicator. No reputable major manufacturer will ever sell a truck that does not meet those requirements, so unless you are undertaking major conversion work such regulations will not be a concern.

> One of the exceptions to the requirement to be aged at least 21 to drive a vehicle over 7.5 tonnes is if the vehicle is a road roller, not powered by steam, and not more than 11.69 tonnes (Lowe and Pidgeon, 2018). This is not something that most of us will in practice ever need to know!

Penalties for failing to comply with regulations can be severe, and can result in a company being banned from operating. If you are considering setting up a vehicle operation, it is important to do your homework well in advance. You will need to comply with numerous regulations, to put in place procedures to ensure that you do so, and to be able to demonstrate your compliance.

In this chapter I will discuss some of the main aspects of legal requirements and compliance.

Driver licensing

> It goes without saying that everyone driving on a public road needs a valid driving licence. As an employer, you have a duty to ensure that your employees have such a licence, that it has not expired, and that they have not been disqualified from driving. This applies to everyone driving a company vehicle, even if it is merely a short trip by an HR assistant to the supermarket because the office has run out of coffee.

You should therefore put in place a system for checking licences: you should accept only seeing an original, not a copy, and ensure details such as date of birth match the company's records. This should be done when the employee first joins, and periodically thereafter, preferably every quarter. You should keep a photocopy or scan to prove that you have made the checks and confirm whether there are any penalty points (the method of doing so in the absence of a paper counterpart was covered in Chapter 10). It should also be a contractual requirement of employment for drivers to notify you of any penalty points or health issues.

If a driver suffers from any one of a large number of health conditions, the DVLA must be informed. For some serious conditions, eg brain tumours, the DVLA must always be notified; for some others, eg depression, this is only required in cases where the ability to drive is affected. A full list can be found at the GOV.UK website under Health Conditions and Driving (UK Government, 2018a).

A 'normal' car licence – technically called a Category B licence – permits the holder to drive a car or van of up to 3.5 tonnes. This includes the weight of any trailer, which must not weigh more than the towing vehicle, or more than 750 kilogrammes. It is possible to get caught out by this, with a towing vehicle such as a transit van within the limit, but the trailer putting the combination over it. Towing a trailer above 750 kilogrammes requires a B + E licence.

Older drivers, ie those who passed their test before 1 January 1997, may also drive vehicles up to 7.5 tonnes using their standard car licence (although there are additional health check requirements on renewal). However, those who passed their test after this date may not do so and require a C1 licence. Perhaps confusingly, whilst the 3.5 tonne limit includes the weight of any trailer, the 7.5 tonne limit does not, so the C1 licence covers a combination up to 8.25 tonnes. Again, a trailer over 750 kilogrammes requires an additional licence, Category C1 + E.

Finally, large goods vehicles require a Category C licence, or C + E if it includes a trailer over 750 kilogrammes. Note that some old licences cover rigid and drawbar combinations only, ie not artics. This is shown on the licence as Restriction 102 (Lowe and Pidgeon, 2018).

There are separate categories for other types of vehicle, such as motorcycles and busses.

Licences to drive larger vehicles are known as vocational licences. The syllabus for vocational driving tests is described in the book *The Official DVSA Guide to Driving Goods Vehicles*, available from bookshops or online booksellers.

Drivers carrying hazardous goods above certain quantity thresholds (eg 1,000 litres for tankers, and zero for explosives) must also pass a specialized qualification known as ADR (an abbreviation of the European Agreement concerning the International Carriage of Dangerous Goods by Road). A special Certificate of Professional Competence (CPC) (see below) is required for drivers transporting livestock more than 65 kilometres (Lowe and Pidgeon, 2018). Whilst not technically a legal requirement, I would recommend that before drivers operate equipment such as cranes they receive training and are tested: an accident caused by an untrained driver would almost certainly be a breach of health and safety regulations.

As discussed previously, there is a shortage of LGV drivers, which is likely to worsen over the medium term. If you are operating a mixed fleet, you may wish to establish a career path such that the best van drivers are selected for C1 training, and the best C1 drivers for category C or C + E training, the costs being paid by the employer. There is sadly a danger that some drivers will adopt a cynical attitude, accept the training, and almost immediately after passing the vocational test leave the company. I would recommend that the costs of training be contractually treated as a loan, so that they can be recouped if a driver takes this approach.

Driver CPC

To raise standards in the industry, it is now compulsory for all professional HGV drivers to hold a Certificate of Professional Competence. This is a level 2 qualification, ie it is intended to be of the same level of difficulty as a GCSE or O level. There are as usual some exceptions, including those who wish to drive an HGV for purely personal use – for example a horse box in connection with a hobby of horse-riding. The initial CPC will in practice be gained at the same time as the vocational licence. Additional theory and practical tests must be passed, which do not purely relate to driving, but cover items such as safe loading of vehicles and documentation.

Ongoing training must be undertaken throughout a driver's career. This can be taken as a minimum of one day each year or as one week every five years. It can only be undertaken at approved training centres, which can be established in-house by larger companies. This training can take the form of 'upskilling', for example, optimizing fuel economy; or of introducing new skills, such as ADR training for the carriage of hazardous goods.

Operator or 'O' licensing

> If you wish to operate vehicles above 3.5 tonnes in connection with your business, you will need an operator licence (O licence). This is the main system for enforcing road traffic legislation, and penalties for operating without a licence are severe. You will need to meet various criteria to be awarded a licence, and it can be removed, for example for persistent disregard of traffic laws.

You do not need an O licence if you are using a 3PL. However, if the 3PL are using your premises as a base, and routinely parking vehicles there overnight, they will need to include your premises as an operating centre on their own licence.

There are a few exceptions where an O licence is not required, including emergency vehicles; those used for funerals; vehicles moving on public roads between two parts of your own premises but not being on the road for more than 6 miles per week; and vehicles not used in connection with a business. There is a separate O licence system for buses and coaches.

However, the vast majority of operators will need an O licence. There were 73,458 in the UK in 2016, and 404,804 registered HGVs (FTA, 2018). This equates to about 5.5 HGVs per licence, so it can be deduced that most are held by small operators.

There are four types:

1. Restricted. This allows you to carry your own goods, but not goods for hire and reward. This is interpreted strictly – even loaning the vehicle for a customer's float in a charity parade would probably be considered a breach. If you ever wish to backload your vehicles (see next chapter) or use them to earn money during a slack period, a restricted licence will not permit you to do so.
2. Standard national. This allows you to carry goods for hire and reward within the UK, and to carry your own goods internationally (Lowe and Pidgeon, 2018)
3. Standard international. Allows you to carry goods for hire and reward both within the UK and internationally.

4 Interim. These may be granted whilst a decision is being made on a permanent licence. They will not be granted in circumstances where you require extra time to meet the requirements of a full licence, nor will they be granted immediately, and they are very unlikely to be issued to anyone who has not previously held an O licence. My recommendation would therefore be to apply in good time for the permanent licence and certainly not to rely on this option.

There are some advantages to a restricted licence. First, some larger companies have a policy of not using their vehicles for anyone else's goods in any circumstances and hold only restricted licences to reinforce this principle. Second, there is no legal requirement to employ a transport manager who holds a CPC, which you must if you apply for a standard licence. However, you will still need to obey a wide range of legal requirements, and possession of a CPC by people employed in key positions may be a good idea in any event. Also, there is a requirement to prove sufficient financial standing, ie you will need to show that you have enough money to maintain the vehicles to proper standards. For a restricted licence the specified amounts for 2018 are £3,100 for the first vehicle and £1,700 for subsequent vehicles, compared to £7,950 and £4,400 for a standard licence (UK Government, 2017).

For any licence, you will need to be of good repute. This in practice means that you must not have committed more than one relevant offence, and not have committed any serious relevant offences. Relevant offences are interpreted quite widely and include not only obvious examples such as contravening driver's hours or dangerous goods regulations, but also contraventions of corporate or employment law and even drug trafficking. My best advice is to obey all laws, and not to employ people who do not.

Operating centres

An operating centre is the place where a vehicle is normally kept, so if there are several places where you do so, you will need several operating centres on your O licence. This may include your customer's or supplier's premises if you routinely keep vehicles there; and if you regularly allow drivers to take vehicles home overnight, then the parking area near their home will need to be approved as an operating centre, following the same procedures of advertising and application to the Traffic Commissioner as any other operating centre.

For each operating centre you will be required to show that there is sufficient off-street parking for all vehicles. You will have to place an advert in a local paper and nearby residents may make representations if they do not approve.

Professional competence: Transport managers

The first thing to make clear is that the CPC for transport managers (TMs) is not the same as the CPC for drivers described earlier.

> To put it simply, if you want to be a nominated transport manager on an O licence, you need to pass an exam to prove that you know what you should be doing.

The exam itself is a level 3 qualification, which means that it is intended to be the same level of difficulty as an A level. There were previously separate national and international CPCs but these have now been merged into a single qualification.

The exam covers both business studies and transport. Full details can be found in *A Study Guide for the Operator Certificate of Professional Competence in Road Freight* by Clive Pidgeon (2016) (published by Kogan Page). There is a separate qualification for passenger service vehicle operators, although much of the content is shared.

Examinations are conducted by the Oxford and Cambridge Royal Society of Arts, which includes multiple choice questions and a case study; and by the Chartered Institute of Logistics and Transport (CILT), which is based on short and long written answers. Training courses are offered by CILT, the Road Haulage Association, Freight Transport Association and various commercial organizations.

Applying for an O licence

> In theory, an application should be made a least seven weeks before operations are due to begin, or nine if you apply by post. In practice, I would always allow extra time, and planning should probably start six months in advance.

You will be required to make undertakings that you will obey all relevant legislation, and that you will take whatever steps are necessary to guarantee that you do so. Before applying, you should make sure that you have all relevant procedures in place to ensure compliance, and will need to satisfy the Traffic Commissioner (TC) that you have done so before the licence is granted. Records must be made available for inspection on request. The areas covered will include:

- drivers hours;
- tachographs;
- overloading;
- speed limits;
- maintenance – minimum frequencies will be specified as part of the O licence;
- defects reported by drivers and action taken as a result (records kept at least 15 months);
- notification to the TC of any major changes, eg a change in maintenance arrangements;
- notification to the TC of any convictions against the company or drivers.

You will need to specify the number of vehicles and trailers to be covered by the O licence. It is wise to leave a 'margin', ie to apply for slightly more vehicles than you expect to use permanently. In an extreme case, if your O licence is for just a single vehicle, and that vehicle is off the road due to accident or breakdown, you will not be able to operate. If, however, your O licence covers two vehicles, you will be able to spot hire another vehicle until your permanent vehicle is repaired.

The application itself can be made by post using form GV79, but it is better to do so online via the Gov.uk website. The licence can be in the name of an individual, a partnership, or a company – whichever is the user of the vehicles. You will need to nominate TMs: if applying by post you will need to complete form TM1 and enclose original Certificates of Professional Competence and proof that they have been engaged to perform the role. A TM can be the licence holder if suitably qualified, an employee or a consultant. They will be required to manage the operation directly, and spend time on site doing so (there are recommended minimum lengths of time per week which increase as the number of vehicles increases). Being a TM in name only is not permitted.

In considering an application, the TC will arrange for a representative to visit the proposed operating centre. They will consider whether the applicant is a fit and proper person and of good repute, financial standing, maintenance arrangements, procedures to ensure compliance, and the suitability of the operating centre.

Certain organizations have a right to make statutory objections to the award of an O licence. They are:

- local councils (but not parish councils);
- planning authorities;
- police;
- certain trade organizations, such as the Road Haulage Association;
- certain trade unions.

These objections are not in practice made lightly, and will be taken very seriously by the TC. If, for example, an applicant attracts an objection because of a poor record of legal compliance when holding a previous O licence, this will almost certainly result in refusal.

Local residents can make representations on environmental grounds – these carry less force than objections but will still be considered seriously. The TC will probably convene a public enquiry to allow them to state their case and consider each representation on its merits.

In most cases, applicants will have prepared thoroughly, ensured that they meet all requirements, and have proposed a suitable operating centre. In these cases, the application will be approved.

If not, the TC has the option to reject the application completely. Alternatively, a smaller number of vehicles than requested may be authorized, or conditions might be applied. Examples include a ban on operations between 22.00 and 06.00 or a stipulation that all vehicles must use the entrance onto a main road rather than the rear entrance into a residential area. There is in theory a right of appeal to a Transport Tribunal in cases of rejection, but my advice would be to ensure that the application is thoroughly prepared to avoid rejection.

Drivers' hours

Vans

It is not widely known that drivers' hours regulations, albeit different to those for HGVs, apply to all goods vehicles: van drivers must obey UK regulations.

The rules are simple – a maximum of 10 hours' driving and a maximum of 11 hours' duty on any day. There is no requirement for record keeping, and in practice breaches of the rules frequently go undetected. There were just 20 convictions in 2014–15, compared to 1,518 for HGV hours offences (Croner-I, 2017). However, employers have a responsibility to ensure that van drivers can complete their day's work within the specified hours, and an accident caused by fatigue due to failure to comply might have serious consequences.

Note also that in Germany vans above 2.8 tonnes must obey the same regulations as for HGVs, and either use a tachograph or keep manual records (Lowe and Pidgeon, 2018).

Large goods vehicles

There are exemptions from the normal rules, including breakdown trucks, Royal Mail vehicles below 7.5 tonnes, mobile libraries, and (for reasons I do not personally understand) vehicles used on some islands such as Arran and the Isle of Wight (Lowe and Pidgeon, 2018).

However, any vehicle over 3.5 tonnes in mainstream commercial use will need to obey drivers' hours regulations. The main rules are:

- Maximum driving hours: nine per day, may be extended to 10 twice per week; 56 per week (four shifts of nine hours plus two of 10 hours); 90 hours per fortnight. Note that these periods are interpreted so as to prevent avoidance. If a driver has been on holiday and driven no hours in week one they may drive 56 in week two. However, in week three the maximum will be 34 hours – you cannot argue that two consecutive busy weeks are parts of separate fortnights. A week in this context is defined as 00.00 Monday to 23.59 Sunday.
- Minimum break of 45 minutes after 4.5 hours driving. This can be divided into a 15 minute and a 30-minute break, but the latter must start at the 4.5 hour point at the latest. Note that on the days driving is extended to 10 hours a total of at least 90 minutes breaks must be taken. Breaks must not be cut short even by a few minutes.
- Daily rest of one 11 hour period, or a three hour period followed by a nine hour period – ie the break must be longer in total if it is split. Daily rest can be reduced to nine hours a maximum of three times each week.
- For double manned vehicles, each driver must have at least nine hours' rest in each 30 hour period. Rest can only be taken in a vehicle whilst it is stationary, so it is not permissible for one driver to rest whilst the other drives, and vice versa. Also, to take rest in a vehicle a driver must be able to lie down on a bunk. If there is only one bunk then to take their rest at the same time one driver must have a bed elsewhere.
- Ferries. In order to take rest on a ferry, drivers must have access to a bunk – since the *Herald of Free Enterprise* disaster they are not permitted to remain in their vehicles during the voyage. Normal 11-hour rest can be interrupted to allow them to drive on and off the ferry, but the interruption must be for no longer than necessary.
- Weekly rest. After a maximum of six shifts, 45 hours' weekly rest must be taken. This can be reduced to 24 hours, but the reduction must be made up as a single continuous period before the end of the third following week.

Note that during rest a driver must be able to freely dispose of their time. If, for example, they are loading the vehicle this will count as other work; sitting in the canteen drinking tea whilst waiting for instructions is deemed a period of availability.

If they take a training course at the weekend this does not count as rest either, even if it is unpaid (Lowe and Pidgeon, 2018).

It is the duty of an employer to schedule a driver's work such that it can be completed within the drivers' hours regulations – if they will have to drive continuously for five hours to meet a delivery time this will not comply. They must also make regular checks of tachograph records (see below) to ensure that rules are being obeyed, and take action where necessary to prevent recurrence of any breach. It is specifically forbidden to make bonus payments that encourage drivers to break the rules.

The same rules apply throughout the EU, and also in Norway, Switzerland and what are known as the AETR countries. This area extends as far as countries such as Turkmenistan and thus gives excellent standardization throughout Europe.

Tachographs

Tachographs have been in use for several decades. A tachograph is a device for recording information relating to the driving of a vehicle and is used to monitor compliance with drivers' hours and other legislation, such as speed limits. Unless exempt, all vehicles over 3.5 tonnes must be fitted with a 'tacho'. Older vehicles, ie those registered before 1 May 2006, will probably still be fitted with their original analogue tacho into which discs must be placed containing hand-written details such as the driver's name. These are still legal, and no doubt it will be some years before they finally disappear from use.

Modern vehicles will carry a digital tacho. These operate using a smart card that is individual to each driver: if they do not have one, they cannot drive an HGV. The smart card will store details of driving, breaks, etc, for a rolling period of 28 days. To obtain a smart card, a driver needs to complete form D777B/DL and send it with the fee (£22 as of 2018) to the DVLA: a company may apply in bulk for up to 25 cards (DVLA, 2018). They are valid for five years, and application for a replacement should be made at least 15 days before expiry.

The tacho unit itself will read the card and store information relating to previous driving of other vehicles, so that a full record of a driver's work is available. On inserting the card, the driver should manually enter activity since the card was last inserted into a tacho. This is usually done using up, down and OK buttons, and observing symbols and numbers on the display. As with so many tasks, this is easy when you know how to do it, but the process may not be completely intuitive, and drivers may require familiarization training if they have not used a particular type of tacho before.

The tacho will have a facility to produce a printout on a till-roll type strip of paper: drivers must carry at least one spare roll in case the one in the tacho runs out, and TMs should ensure that they do not run out of stock. One possible source of confusion is that whilst local time may be shown on the display screen, the printout will show times in Universal Time Coordinated (UTC), ie Greenwich Mean Time.

A printout should be generated by the driver at the end of each shift, and must be produced on request from a police officer or DVSA traffic examiner at the roadside.

Tachograph analysis

The old-style tacho discs were difficult to read accurately, and most companies had to employ outside agencies to analyse them. However, one of the advantages of digital tachos is that you can purchase software and hardware to conduct your own analysis. You will require a company smart card (also obtainable from the DVLA, by completing form D779B) to conduct the download.

The software will permit you to produce reports that summarize the records. A fictional example is shown in Figure 11.1: if any real driver shares one of these names this is purely coincidental. It can be seen that downloads have been made in respect of 12 drivers. A smart card logo shows that data is available for that driver for that day; a truck logo that they drove a hired vehicle; R that this was a rest day, etc. On a colour screen red flags indicate an area of concern.

In Figure 11.2 we can see a detailed breakdown of the driving record of the fictional Mr Sandomierz. He has committed an infringement, ie he has failed to take a sufficient break. Action with respect to any infringements such as this must be taken and recorded: in cases of serious or repeated offending this action will almost certainly result in dismissal of the guilty driver.

Compliance monitoring

As I stated earlier, you will have been required to give an undertaking to comply with all the requirements of your operator's licence. Tachograph analysis is one part of this, but you will also need to show, for example, that vehicles have been maintained in accordance with the time and mileage intervals specified, and that everyone involved in the operation has the necessary skills and qualifications to perform their role.

It will assist the process if you make use of a suitable software package, such as Silk Thread®, which is specifically designed to assist in the monitoring of compliance. The system requires answers to specific compliance-related questions, and then conducts a gap analysis to highlight areas where further action is needed. In the fictional example in Figure 11.3, it can be seen that the Wakefield depot has completed all necessary entries for the period, but the yellow flags (which show as a lighter shade of grey) indicate two non-critical areas of concern. Drilling down into the system will reveal the details: perhaps, for example, a driver was convicted of mobile phone use but no action has yet been taken to advise the Traffic Commissioner. The system can also produce diary reminders and action lists to ensure that necessary actions are not forgotten.

Figure 11.1 Summary report of downloaded information relating to 12 fictional drivers

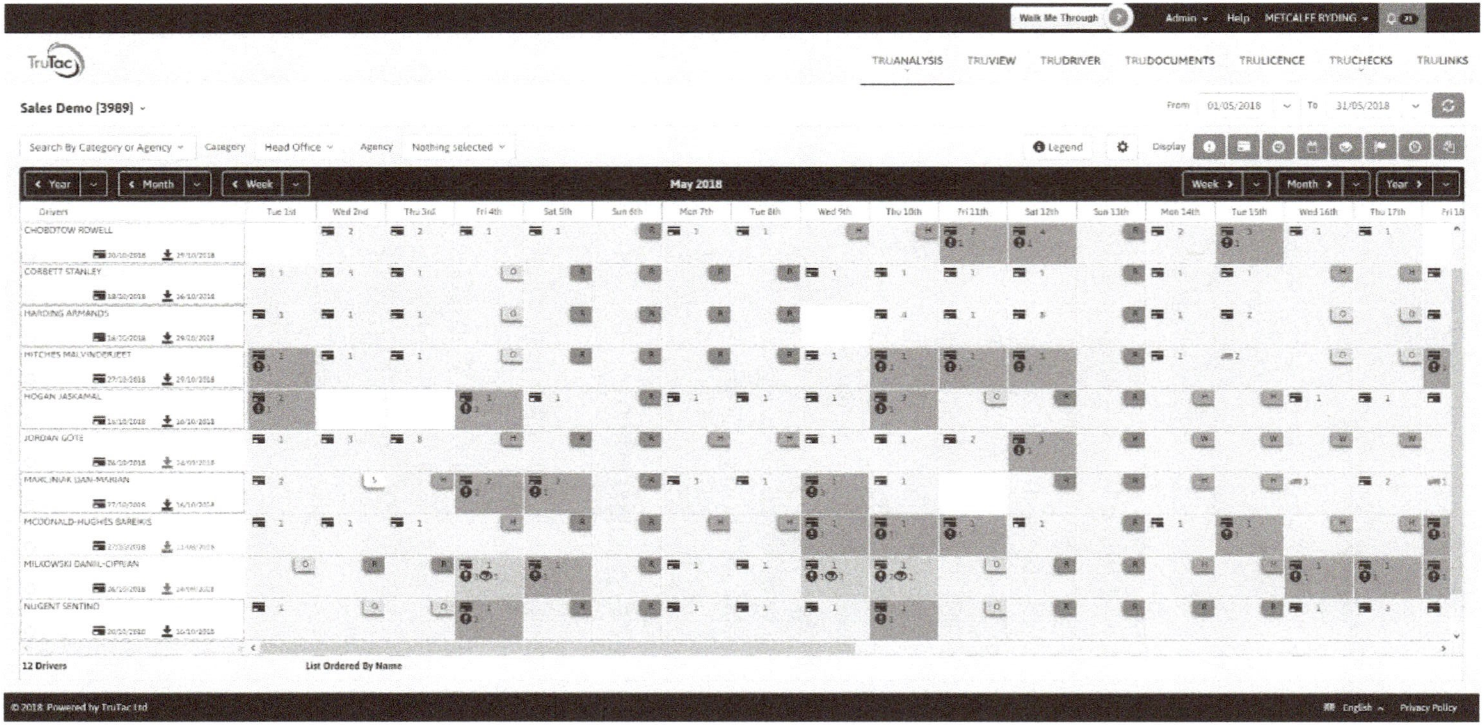

Walk Me Through | Admin | Help | METCALFE RYDING

TruTac

TRUANALYSIS | TRUVIEW | TRUDRIVER | TRUDOCUMENTS | TRULICENCE | TRUCHECKS | TRULINKS

Sales Demo [3989]

From 01/05/2018 To 31/05/2018

Search By Category or Agency | Category | Head Office | Agency | Nothing selected | Legend | Display

Year | Month | Week | May 2018 | Week | Month | Year

Drivers	Tue 1st	Wed 2nd	Thu 3rd	Fri 4th	Sat 5th	Sun 6th	Mon 7th	Tue 8th	Wed 9th	Thu 10th	Fri 11th	Sat 12th	Sun 13th	Mon 14th	Tue 15th	Wed 16th	Thu 17th	Fri 18
CHOBOTOW ROWELL																		
CORBETT STANLEY																		
HARDING ARMANDS																		
HITCHES MALVINDERJEET																		
HOGAN JASKAMAL																		
JORDAN GOTE																		
MARCINIAK DAN-MARIAN																		
MCDONALD-HUGHES BAREIKIS																		
MILKOWSKI DANIL-CIPRIAN																		
NUGENT SENTINO																		

12 Drivers | List Ordered By Name

© 2018 Powered by TruTac Ltd | English | Privacy Policy

SOURCE Reproduced by kind permission of TruTac Ltd

Figure 11.2 Fictional detailed driving record from tachograph download

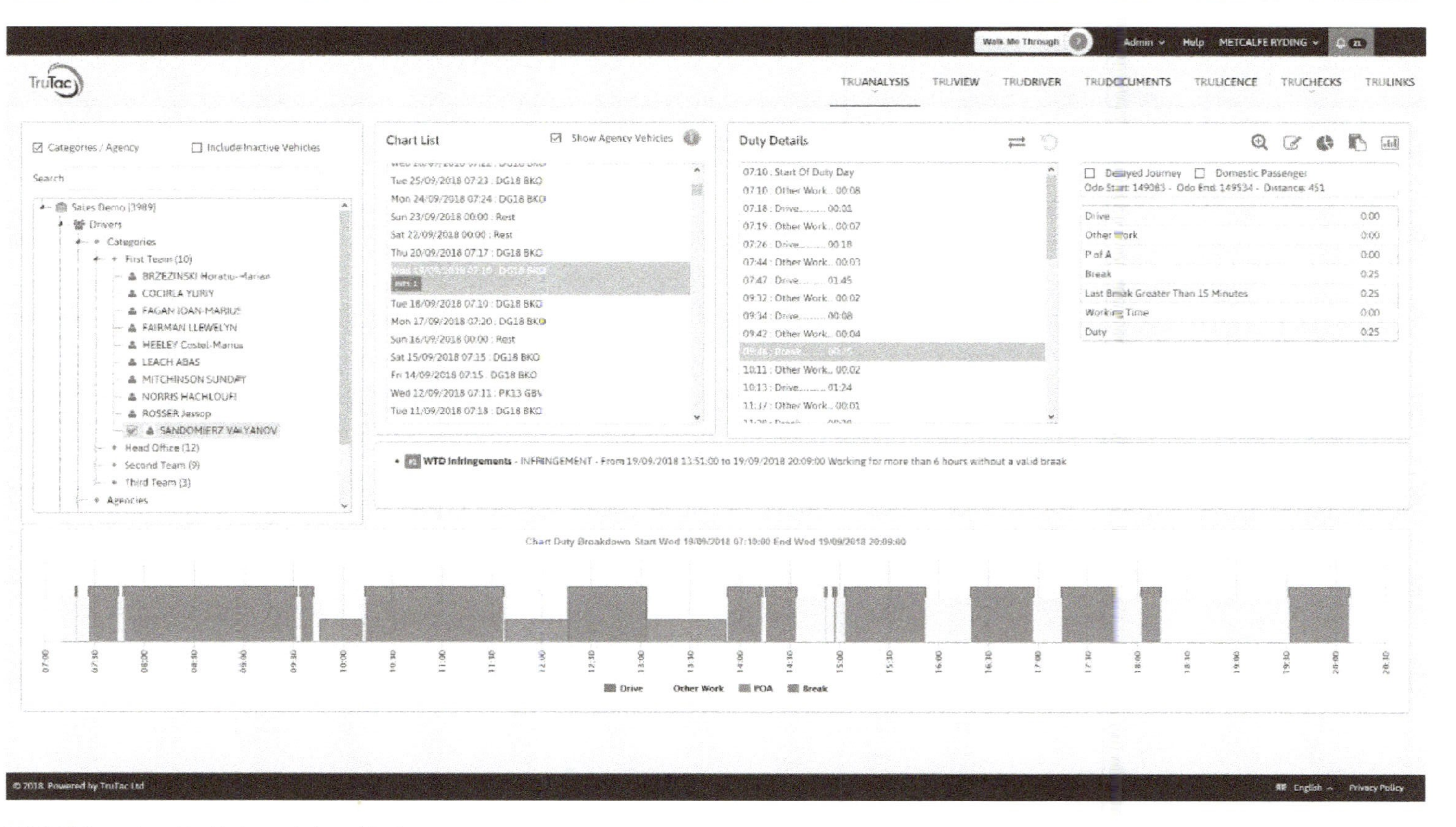

SOURCE Reproduced by kind permission of TruTac Lt

Figure 11.3 Example of a compliance status dashboard summary from the Silk Thread® system

Topic Dashboard for audit started 16/06/2017 10:07

Started:	Last Modified:	Completed:	Expiry:
16/06/2017	24/06/2017	24/06/2017	11/08/2017

Topic	Status	
O Licence:	100%	View Answers
Duties of CPC holder:	100%	View Answers
Management and Supervision:	100%	View Answers
Driver Profiles:	100%	View Answers
Speeding:	100%	View Answers
Maintenance:	100%	View Answers
Loads & Axle Weights:	100%	View Answers
Skills & Competence:	100%	View Answers
Managing driver hours:	100%	View Answers
Working Time Directive:	100%	View Answers
Managing Agency Staff:	100%	View Answers

SOURCE Reproduced by kind permission of Labyrinth Logistics Consulting Ltd

Laws in non-UK countries

Many elements of road traffic law, such as drivers' hours, are harmonized throughout the EU and other European countries.

However, individual countries do have their own specific laws. Examples include:

- France:
  - Ban on Sunday and Public Holiday driving for HGVs.
  - Speed limits reduced in wet weather.
- Belgium:
  - Overtaking prohibited during rush hour.
  - Restrictions on use of cruise control.

- Scandinavia:
  - Dipped headlights must be used at all times, even in bright sunshine.
- Germany:
  - An amber flashing light must be carried in case of breakdown.
  - Drivers delivering or collecting in the country must be paid at least the German minimum wage (Lowe and Pidgeon, 2018).
- Spain:
  - Drivers who wear glasses must be carry a spare pair in case their main pair is lost or broken. This would seem to me to be a sensible precaution in all countries.

There are also requirements in several countries to carry spare bulbs, warning triangles, and high visibility clothing. Again, these would seem useful items for all vehicles to carry.

My advice in all cases is to familiarize yourself with the laws of any country in which your drivers will be operating. Some police forces target overseas drivers, and penalties can be severe – ignorance of the law will not be accepted as an excuse.

Further reading

In this chapter I have provided a summary of some of the main legal requirements for vehicle operations, with particular emphasis on driver licencing, operator licencing, and drivers' hours regulations. Transport law is a very complex subject, and if you would like to study it in greater detail there are books available to do so. These include the *Transport Manager's and Operator's Handbook* by Lowe and Pidgeon (2018), which extends to 743 pages, much of it devoted to transport law; and the Freight Transport Association's annual *FTA Yearbook of Road Transport Law*. Those operating internationally may find the FTA's *European Road Freight Guide* to be a useful aid.

Conclusion

Over the course of the last three chapters I have discussed what sort of vehicle you are likely to need; the costs of running vehicles, and how they can be reduced; and the legal requirements for vehicle operations.

In the next chapter, I will discuss how best to operate those vehicles on a day-to-day basis, and how to route them to best effect.

Vehicle routing and networks 12

In Chapter 10 I discussed the costs of road vehicle operations, and how they can in some cases be mitigated. However, the most important aspect in a cost-effective operation is to use the vehicles efficiently, routing them in the best possible manner.

Today, there are software packages available to assist in some parts of the process, which I will discuss later. However, I will first look at the various methods available for vehicle operations. Most comments will be equally applicable to in-house or 3PL situations.

Planning delivery bookings

The first point is that some companies make efficient vehicle operations unnecessarily difficult for themselves. Bookings for delivery slots are made at short notice, and with little to no thought of the transport operation. It may even be better for bookings to be made by the logistics department rather than the sales department. Some of the simpler steps to take include:

- Book timed deliveries as far in advance as possible. This will give you the widest choice, and ample opportunity to schedule other deliveries around the slot time.
- Book deliveries to the same geographical area at times that allow them to be made on the same vehicle.
- Avoid rush hour times, which would result in driving through heavily congested traffic – deliveries to inner city areas in the rush hour are not recommended.
- Avoid making unnecessary promises – if a customer is happy to receive a delivery at any time during the day, do not insist on giving them a specific time slot.
- Remember past difficulties and take them into account. If a customer takes a long time to unload your vehicle, consider this in future bookings.
- Do not book deliveries to your full capacity too far in advance, leaving no spare capacity for urgent requirements.
- Worse still, do not book deliveries to exceed your full capacity!

Full load and single drop operations

If you are selling goods in sufficient quantities, full load artic deliveries will be the most efficient method. If you can persuade your customer to order in the appropriate quantities, this will minimize your costs and enable you to offer the best price to the customer. If you are using a 3PL, this will be reflected in their rates to you, especially for international movements.

Other circumstances in which you may decide to make a single delivery with a vehicle include:

- The customer insists that only their goods are loaded on the delivery vehicle – this is not always a sensible restriction, but the customer is always the customer.
- The goods are high value, and/or vulnerable to damage, such as fine arts.
- There are security requirements, especially in the defence industry.
- The delivery is urgent.
- The driver needs to spend a lot of time installing the items, such as solar panels.
- There is a lack of other freight to make up a viable route.

The last of these may be unavoidable at times, and at others the most economical option even though it is unattractive.

For example, let us assume that you are based in London, and tomorrow have two-half loads to deliver in the West Midlands, and one-half-load to deliver in rural Norfolk:

- Your first thought will be to deliver the first two on your own vehicle and subcontract the rural delivery to a local 3PL. However, this job will be unattractive to a London-based haulier who has nothing else to deliver in the area and will have to return empty.
- Their price will reflect this: it may, indeed, be more than their price for the two drops in the Midlands.
- The best solution for you may therefore be to send your own vehicle to Norfolk and subcontract the other jobs. If the rural deliveries are going to be regular, the long-term solution may be to find an East Anglian 3PL who will be willing to backload, but this option may not immediately be available.

Part loads

Not all deliveries fall into the category of either full loads or of small consignments; some fall in between the two. These are known as part loads.

Many international 3PL operators will have their own definition of a part load, maybe any consignment over 1,500 kilogrammes or any over 3 tonnes (in practice this will be a reflection of how much spare capacity they have in their shared user network). If you have, say, six or 10 pallets for delivery to Milan, these can be shipped as a part load, ie a vehicle will collect your consignment and deliver it directly to your customer – the expression 'on wheels' is sometimes heard. It is the responsibility of the 3PL to find other goods to fill the remaining space on the vehicle.

It is worth noting first that some 3PLs have been known to price large part loads (eg 15 pallets) higher than a full load. If it is cost-effective to book a full load trailer then you should do so.

Second, if you have several part load deliveries to make, look for opportunities to combine them on a single vehicle. It may be cheaper to book a full load and pay a surcharge for an extra drop than for two part loads. For 10 pallets from the UK to each of Milan and Turin this will almost certainly be the case; for six pallets to each of Milan, Rome and Naples it will usually be.

Finally, for domestic movements, a single drop on a smaller vehicle will be more economical for shorter distances. If you have 10 pallets to deliver from London to Northampton, then a dedicated 18 tonne truck will probably be the best solution; from London to Glasgow, this will be best shipped as a part load on an artic. Determining the best solution for your particular business is a process of considering each case on its merits, and retaining this information for future reference.

Return loads or backhaul

If you have made a long-distance delivery, perhaps from London to Glasgow, the idea of returning empty is unappealing from both the economic and environmental viewpoints. If the vehicle can be loaded in both directions, there are obvious benefits.

If you are using a 3PL, then in narrow terms the risk and cost of not finding a backload, and the financial rewards for doing so, rest with them (although their pricing will obviously reflect this). However, if you are using your own vehicle, you must deal with the problem.

The first source of potential backloads should be your own business. If, for example, you have warehouses in Doncaster and Bristol, an artic from Doncaster that has made a delivery to Cardiff could collect a load from Bristol for delivery to Leeds. Inter-depot trunking is also a common opportunity to backload. You may also have items to collect from your customers, such as empty stillages if you are serving the automotive industry, or unsold goods if you have supplied a retailer on a sale or return basis. There may also be returned goods – TV shopping channels tend to have a 'No questions asked' returns policy for their customers, which are then passed back to their supplier. There is, however, a word of warning that collecting small quantities of returns with a large vehicle may be inefficient – one pallet on an artic that prevents a full load being carried will be uneconomic. It would probably be better to use the artic for a full load and return the pallet via a pallet network (see below).

Alternatively, if you are using a pallet hire service such as CHEP, you could collect empty pallets using your own vehicle and thereby save a delivery charge.

You might also consider changing the terms of purchase with your suppliers. If they currently deliver, but you regularly deliver to your customers in the same area, it should be possible to negotiate a price reduction if you collect, effectively paying for the return leg. You will be able to do this under a restricted O licence, as the goods are yours at the point you collect, but if you simply invoice the supplier for the transport the goods are not yours at the point of carriage and you will need a standard O licence. This is an important technical point to remember.

The above options are worth considering even if you are using a 3PL. If they can be sure of a backload this should be significantly cheaper than sourcing two single legs.

Finally, you could seek a return load elsewhere. You may be able to transport goods on behalf of your customer, delivering to their customers. There are also websites where loads are advertised, to be matched with trucks with spare capacity. However, if you do take this approach, please remember that you will need a standard O licence, appropriate insurance, and the establishment to handle matters such as credit control. There is also a risk of vehicles being delayed at unloading points and not therefore available when you need them. If you are considering this direction, it is not a decision to be taken lightly.

It is often said by opponents of the roadfreight industry that the number of trucks on the road could be reduced if all vehicles were backloaded. At first sight they have a point – 30.2 per cent HGV empty running was reported in 2016 (FTA, 2018). However, life is not as simple as that.

There are numerous reasons why backloading may be impossible or uneconomic, for example:

- Especially for short distances, the need to relocate to another location to obtain the backload, then relocate again to your own premises may render the idea too costly.

- Backloads may not be available on a particular route, especially between rural areas.
- Material flows are not always balanced, especially for specialized vehicles, eg:
  - Devon and Cornwall consume all sorts of products, but produce little, so most curtainsiders and every petrol tanker must return empty.
  - 10 per cent of UK car sales are in Scotland, but all factories and most ports for car imports are in England, so it is almost impossible to obtain a backload for a car transporter out of Scotland.
  - The UK imports about 50 per cent more from Germany than it exports, 100 per cent more in the case of Spain (ONS, 2018). There is therefore a big imbalance in the flow of freight.
- There is no real market for backloads for small vehicles. If you run an urgent order in a van, it will almost certainly have to come back empty.
- The risk of losing availability of a vehicle due to delays on a backload may be too great, especially for perishable goods or just in time deliveries. One major retailer routinely turns down backload opportunities because of this risk.

Taking all of this into account, there will unfortunately always be a need for some empty running.

Utilizing the backload option for your products

Many companies have one or a few customers in difficult geographical locations – perhaps rural Lincolnshire or mid Wales. It will be expensive to send out your own vehicle, and unattractive to a 3PL who will charge accordingly. One possible solution is to find a haulier based in that area who would be prepared to collect from you and deliver to your customer on a backload basis. You should of course vet the haulier, to satisfy yourself as to their suitability, but this may be the most cost-effective routing.

Some companies use the backloading option more heavily, utilizing a range of contractors in different areas. If you use the services of a 4PL (see Chapter 2) they are likely to do so quite extensively.

Night operations and double-shifting

In many situations, operations are restricted to daytime hours by their very nature. If for example you are delivering to small retail stores or other SMEs, most will be open only from about 09.00 to 18.00.

If you can operate by night, this will have advantages. First, traffic conditions are likely to be much less busy, especially in urban areas, improving productivity and reducing delays. Also, there are clear financial advantages to double-shifting a vehicle. Based on the figures in Table 10.1, the standing cost for a 44-tonne artic is a little over £50,000 per year (excluding driver's wages). If it can be double-shifted, obtaining twice the productivity from what is at the end of the day an expensive asset, this will far outweigh the night-shift premium to be paid to the driver.

This principle is worth remembering when negotiating with a 3PL. Some will actually try to charge a higher price for a second shift, justified by night-time wage rates, effectively charging you twice for the standing cost of the vehicle. I would not advise you to accept this.

Opportunities for night-time deliveries can be found with delivery destinations such as major retail distribution centres, which are usually open 24 hours per day, and may be very happy to give you an 02.00 delivery slot. Even if the warehouse is closed at night, there is likely to be a security presence, and if they are operating a stand trailer system they will probably permit night-time changeovers. Also, you may be able to arrange inter-depot trunking within your own organization, again combined with a stand trailer system if appropriate.

Finally, some networks are set up for night-time delivery into lock-up facilities. This method is used by several major car manufacturers. Parts urgently required to repair vehicles off the road due to breakdown (known as VOR) can usually be ordered up until late in the afternoon. They are then picked and dispatched from a warehouse in the Midlands and deliveries of parts ranging from a screw to an engine are made into a lock-up at each of several hundred dealers every night, using a fleet of dedicated vans operating from regional centres. Guaranteed deliveries as far afield as Plymouth and Glasgow by 08.00 the following day can usually be achieved by such networks.

A variation on this idea is that some companies make night-time deliveries into lock-ups for collection by their own engineers, for installation or repairs.

One caveat is that some urban areas have restrictions on night-time movements by HGVs. This includes London, where movements between 21.00 and 07.00 Monday to Friday, or between 13.00 Saturday and 07.00 Monday, are restricted to certain main routes unless permission is given (TfL, 2018b).

Multi-drop deliveries

The above sections have mainly been concerned with full and part loads. However, many deliveries will consist only of a small number of pallets, parcels, drums, reels, or other small units for each delivery point. Unless there is a good reason to use a dedicated vehicle for each delivery (see above) then these will require what are known as multi-drop deliveries, ie a vehicle will make numerous deliveries along its route.

One option would be to arrange collection by a 3PL, who will make deliveries via a shared user network; at the other extreme you may have sufficient business to establish your own network. Each of these will be discussed below.

However, a common policy is to put most goods through a network but make deliveries in your own area using a small number of dedicated vehicles – maybe only one. If you do not wish to take on the legal responsibilities of an operator's licence, a 3PL could operate a single vehicle on your behalf. For this type of work, a small local haulier, or maybe even an owner-driver, would most likely be a better option than a major operator. An agreement to guarantee work for a 7.5 or 18 tonner for a minimum of, say, 230 days per year will be attractive to such a business, and likely to attract a favourable rate. If you do not have sufficient traffic for this, you could agree to pay a daily rate for such a vehicle on your busy days, subject to an agreed maximum mileage, differences above and below the agreed figure to be netted off on a monthly basis. It is extremely difficult to generalize, and you should make your own calculations, but daily hire of an 18 tonner will probably become viable if you will be delivering more than about six or seven pallets during the day.

Network distribution

Large businesses

In Chapter 3, I asked the following question: 'How many storage warehouses do I need?'

The conclusion was 'Probably only one.' This is not to say that you should not have more locations for transport operations, possibly with cross-dock warehouses. If you have sufficient business to operate a significant fleet making multi-drop deliveries in an area, it may be cost-effective to trunk to a regional centre, and cross-dock onto local delivery vehicles. The network depots could be in-house, subcontracted to a 3PL, or a combination of the two.

However, the vast majority of businesses will have insufficient traffic to make this a dedicated network such as this a cost-effective solution.

Smaller businesses

If you fall into this category, the solution will probably be to use a shared user network. Within the UK, many networks specialize in the delivery of either parcels or pallets, and some will insist that other types of packages, such as drums of liquid, are palletized for ease of handling.

For international road haulage, shared user shipments are known as groupage, and a vehicle may carry a mixture of pallets, parcels and myriad other items. Major forwarders will publish details of days on which export and import groupage trailers depart to and from various destinations, and the expected transit times. Davies Turner, for example, have daily departures to France, but only a Friday departure to Greece. Expected transit times are two to three working days to Northern Spain, but six to eight days to Turkey (Davies Turner, 2018).

Some shared user networks are operated on a franchise system, or a major operator may use mainly in-house resources plus some subcontractors.

A network that uses a slightly different business model (the only member-owned UK network) is Palletline, which I will now describe in more detail as an example of a shared user network (Figure 12.1). They own a main hub in Birmingham, with regional hubs in Glasgow, Manchester, Leicester, Coventry, London and Swindon; the 78 members operate a total of 96 depots. They each collect freight from their own area during the day; consolidate at their depot; trunk it into the hubs during the night; collect pallets from the hub for their own respective area; trunk back to their local depot; and distribute from that depot to delivery points throughout their zone.

Figure 12.1 The Palletline Hub in operation

All goods are palletized, albeit in a wide variety of shapes and sizes. Vehicles are unloaded in the central area, and pallets sorted by destination in the painted lanes. They will then be loaded onto outbound vehicles via the docks at the sides of the building

SOURCE Reproduced by kind permission of Palletline

An average of 18,000 pallets move through the system each night. Most consignments are between one and three pallets, with an average of 1.46 as at 2018. Their own network is based on kerbside delivery with a tail-lift – goods delivered include TVs on pallets for residential properties. They will unwrap pallets on arrival and take away packaging for a fee, and also make a surcharge for pallets over 750 kilogrammes to allow for an electric pallet truck. They offer timed deliveries (+/- 15 minutes), and for destinations such as supermarket RDCs they have agreed slot times once or twice per day on which they make consolidated deliveries.

The major parcels networks operate in a similar way, and may use roll cages or other forms of consolidation to ease handling at their hubs. An international groupage consignment may be shipped by several hubs – few will operate direct groupage trailers to every country, and certainly not to every region of every country.

When selecting a network, it is important to ensure that the delivery service you require, and which your customers expect, will be provided. Not all networks provide a tail-lift, or they may charge a supplement to do so, but if you are delivering to business premises with a forklift this is not a problem. Items such as furniture may need to be carried into a house, perhaps upstairs in the case of a bed or wardrobe. Two-man deliveries will be required in such cases, and there will clearly be a major customer service issue if an elderly lady is expecting such assistance, but the carrier does not provide it.

Multi-drop vehicle routing

The optimum routing of vehicles is one of the most under-rated skills in the transport industry. A first-rate vehicle router can save the company large amounts of money, by minimizing vehicle mileage and reducing waste, especially in a multi-drop situation. Efficient routing also helps the environment by reducing fuel usage and emissions, and customer service will be improved by eliminating late deliveries.

If you choose to route your own vehicles, you will find the exercise easier if you have an innate talent for maths, and a good geographical knowledge – not just of delivery locations, but which roads to use and which to avoid at various times. My approach was always to start with the fixed requirements, such as timed deliveries. Routes can then be built around these. Circular routes from base for all vehicles with evenly spread deliveries are attractive, but the 'lollipop' method will in many cases be more efficient – a vehicle has a straight run to the start of its delivery area, preferably by motorway, and then makes deliveries within that area. Alternatively, the most urgent delivery may be at the most distant point, so a truck has a long run to its first drop, and then makes the remainder of its drops on its return journey.

A good multi-vehicle plan may well have vehicles following all these types of route on a given day. There should be no fixed rules when planning, except for always operating legally. If it is most efficient for one vehicle to have twice as many drops as another, then so be it.

Vehicle routing is not easy, and it can be time-consuming. However, a good route planner can be extremely valuable to an organization.

Routing software

If you do not have a talented route planner, and/or do not have the time to allocate to the task, then I would recommend a vehicle routing software package. Some comprehensive ERP systems, such as SAP, offer a route optimization module, but for an SME a stand-alone system such as Maxoptra could be considered.

Such a system will allow you to upload orders manually, with a separate data entry for each; by uploading an Excel spreadsheet, which could be downloaded from an order system and emailed to the transport manager; or directly via an application programming interface (API). Details of your depots and available vehicles and drivers will have been pre-entered and can be edited as necessary. When all orders have been uploaded, they can be displayed on a map. Routing can be done semi-automatically, by dragging and dropping, or at the click of a button the system will optimize the drops and allocate routes to vehicles. The system will take into account driver's hours regulations.

A good system will then support the sending of that information to drivers, to be read on a smart phone or tablet, and ideally there will be an interface to the vehicle's satellite navigation system. Combined with global positioning system (GPS) tracking, it will be possible to provide real time information about the progress of a vehicle along its route, comparing scheduled and actual times of deliveries. Updates can be sent to a customer giving a latest estimated arrival time, and it is even possible to mount a large-screen TV on the office wall so that staff can see the whereabouts of any vehicle at any time.

Figure 12.2 shows a screenshot of five fictional routes operating out of a depot in Uckfield. The individual routes follow a variety of paths, including one that is circular; a lollipop pattern via Tunbridge Wells (driven by Quincy Quinley) with a straight run to just North of the Dartford crossing and four deliveries around East London; and a route that makes the most distant delivery first, with numerous drops on the return leg. It can also be seen from the box on the right that after returning to the depot, Mr Quinley will reload and make two further delivery runs during the day.

Figure 12.3 shows a different screenshot from the same system, with a more detailed breakdown of each driver's route showing the individual deliveries.

Figure 12.2 Screenshot from a vehicle routing system, showing a map display of five routes

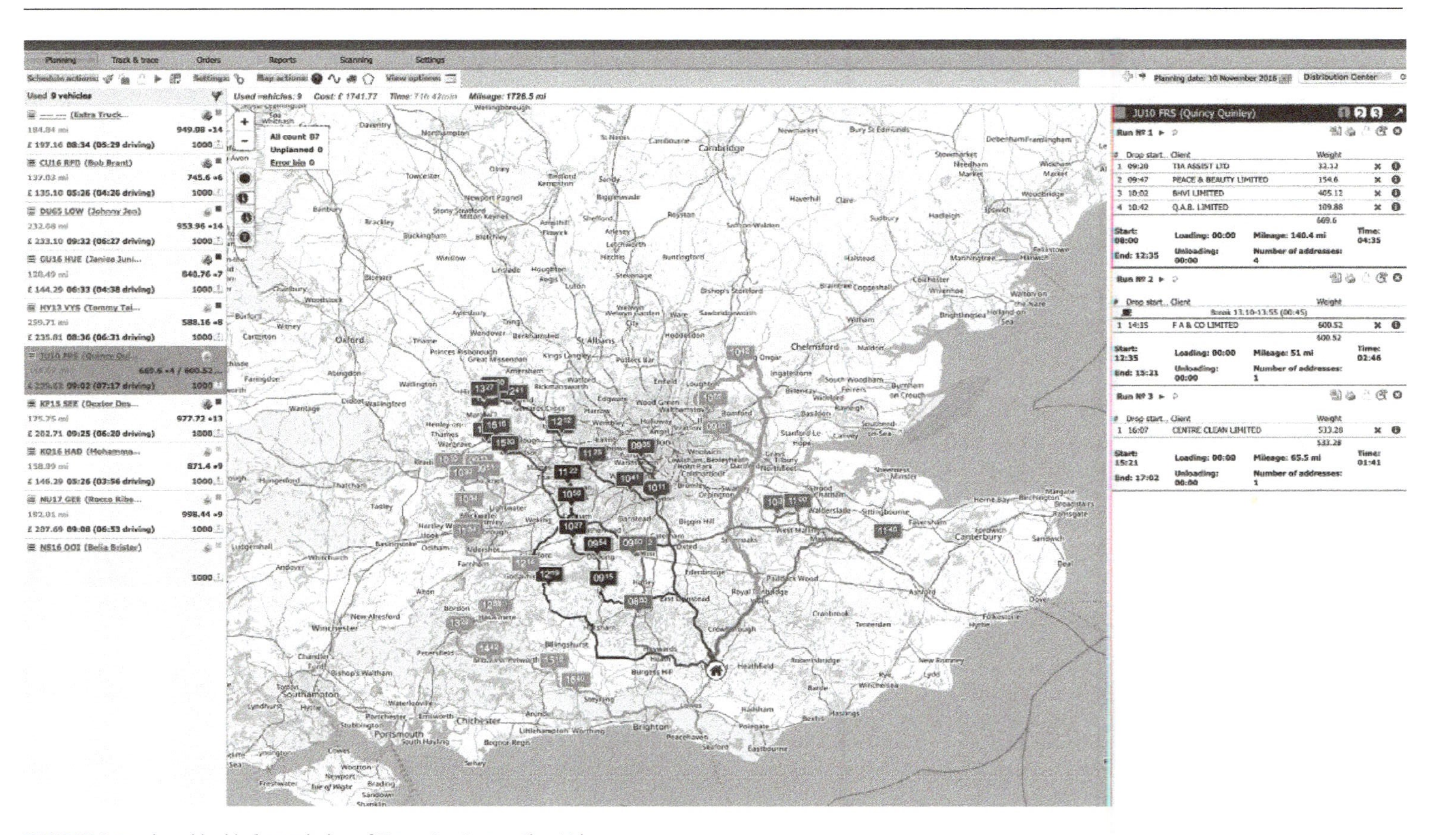

SOURCE Reproduced by kind permission of Maxoptra Corporation Ltd

Figure 12.3 Screenshot showing more detailed information for delivery routes

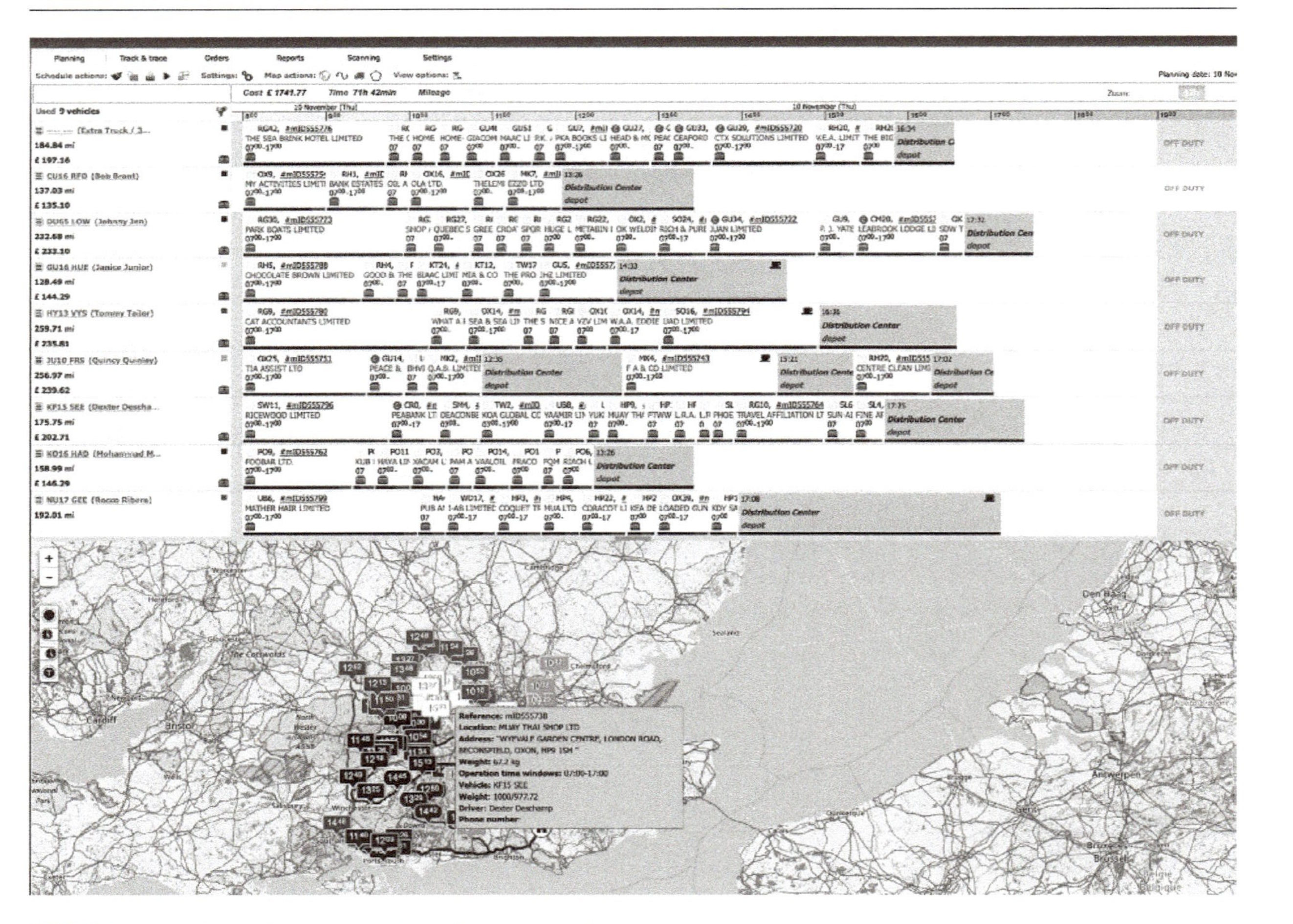

SOURCE Reproduced by kind permission of Maxoptra Corporation Ltd

Proof of delivery tracking

One further benefit of modern telemetry is that GPS locations can be recorded when obtaining a proof of delivery (POD) signature. It is not unknown for items to be mis-delivered, nor is it unknown for them to be correctly delivered and then lost within the customer organization. In the event of a dispute, being able to precisely state where the signature was made can be a valuable tool in establishing the true facts.

Beware of the 'unload round the corner' scam! This works by approaching a driver, often whilst queuing in the road waiting to enter premises for unloading, and announcing that the warehouse is full (or some other excuse), and that the vehicle will therefore need to be unloaded elsewhere. The driver then complies. Often the first hint of a problem comes a few hours later when the customer calls to say the goods have not arrived, and on investigation the story is revealed. By this time the goods and all people involved are long gone and will never be seen again. Drivers should be instructed to refer all such requests to their supervisor, and not to comply unless the supervisor has received written confirmation from a trusted source confirming the revised delivery address.

Conclusion

In this chapter I have described the various options for making deliveries – full and part loads, and dedicated and shared user networks. I have also discussed efficient vehicle routing, and possible use of software to assist this process.

This should be considered alongside the other chapters covering the type of vehicles you might need; the cost of operating them; and the legal obligations in doing so. I have also included a short chapter on the specialized subject of abnormal loads, which many logistics users will encounter occasionally. I hope that together these chapters have provided a comprehensive introduction to the use of road transport in the freight logistics context.

Road transport is not of course the only mode available, and I will now move on to look at some of the others, beginning with railfreight.

13 Railfreight and other inland transport

So far, I have talked extensively about roadfreight, which is by far the most common form of transport used throughout the world. However, it is not the only option. In this chapter I will discuss other modes available for inland transport: railways, inland waterways, and – briefly – fixed installations, ie pipelines and long-distance conveyors.

Although this book is written in the UK, most of the contents apply equally in other countries, whether considering warehousing, roadfreight, seafreight, or many other topics. There are of course some differences, but these tend to be of detail rather than fundamental. However, with respect to railfreight the situation in the UK differs significantly in many respects from that found elsewhere.

This arises because in the UK the viability of railfreight is adversely affected not only by shorter distances compared with many other countries, but also by the age of some of the infrastructure. Higher platforms and lower bridges and tunnels than many other European countries restrict the cross-section of wagons, and the lengths of passing loops (which allow a slower train to move aside so that a faster train can pass), signalling and other factors limit the length of trains. Infrastructure improvements have taken place to address the problem: trains can be operated at up to 75 per cent greater length than previously, which has driven down costs. Similarly, rail transport of high cube containers from the ports of Southampton and Felixstowe is now possible on selected routes (technically, those improved to the W10 gauge). However, this still permits only a maximum width of 2.5 metres and height of 2.9 metres, compared with, for example, 3.15 and 4.65 metres in Germany. In the US, the Schnabel WECX 801 railcar rises 5.5 metres above track level (Railway Age, 2012). This is an illustration of the limitations of even modernized UK infrastructure relative to that elsewhere.

Similarly, most British canals were built with locks just 2.13 metres wide and 22 metres long, limiting barge capacity to about 35 tonnes, a stark contrast to the great waterways of Europe.

Further afield, differences are even more dramatic. In 2001 in Australia a 7.3 kilometre long train carried 82,000 tonnes of iron ore (Railway Gazette, 2001); and ships of up to 305 metres in length sail the Great Lakes of North America.

I will therefore attempt to generalize as far as possible, but it should be borne in mind that economic realities will vary between different geographical areas.

Railfreight

The first point I should make about railfreight is that it is not in most cases a simple alternative to roadfreight. For most applications, such as door-to-door delivery of a single pallet or a parcel, roadfreight is the obvious, and in many cases the only, choice.

Having said this, in 2012 railfreight accounted for 11.2 per cent of UK land freight movements, and 45.9 per cent in Switzerland (European Commission, 2015).

There are advantages:

- To the environment. Each tonne of freight transported by rail creates just 24 per cent of the carbon emissions for a similar road movement (Network Rail, 2018).
- In reducing road congestion. Each train takes on average 76 HGVs off of the road, equating to 1.66 million fewer HGV kilometres each year in the UK (Network Rail, 2018). This in turn reduces deaths and injuries on the roads.
- In reliability. This may be surprising to some, given the publicity often given to passenger train delays. The Rail Delivery Group (2018) reports delays of only approximately five seconds per mile for railfreight, compared to 10 on motorways, 50 on local A roads and 80 seconds per mile on city centre A roads. The overall result is that rail is about twice as reliable as road in terms of on-time delivery.
- In cost. It can be economically competitive, and in some cases much cheaper, than roadfreight.

The main disadvantage is a lack of flexibility. On some routes there may be only one departure per day, whereas a road vehicle can leave at any time over the 24-hour period.

Capacity is also finite. A great many lines have been closed since the heyday of the UK railways, which has often resulted in the loss of alternative routes that could have been used for freight. For example, to relieve pressure on the congested East Coast Main Line, many freight trains are routed from Peterborough to Doncaster via

Spalding and Lincoln. The closure of the March to Spalding line has prevented extending this routing to London via Ely and Cambridge. Whilst such decisions may have been justified when they were made, they have a serious negative impact today.

In the UK, priority tends to be given to passenger trains, especially on busy lines. However, this is not always the case elsewhere. For example, in New Zealand passenger services are restricted to commuter trains serving the Auckland and Wellington areas; and tourist trains on four scenic routes, none with a frequency of more than one train in each direction per day. All other usage of the network is for freight. Freight is also the dominant traffic on long distance routes in the USA and Canada.

The other disadvantage is that although railfreight can in some cases be the cheapest alternative, this is not so for the majority of shipments.

With this in mind, there are six main sets of circumstances in which railfreight might be considered as an option:

- rolling highway;
- bulk haulage;
- intermodal transport;
- single wagon service;
- freight by passenger train;
- fast freight services, a new innovation.

Rolling highway or rolling motorway

This will be most familiar to those in the UK as the system used for passing through the Channel Tunnel. Vans up to 3.5 tonnes can use the same shuttles as cars, but HGVs have their own dedicated service.

The vehicle is driven onto the shuttle train from the terminal platform, and parked as instructed by staff. The main differences between car and HGV shuttle wagons are that the former are enclosed, and the driver and passengers remain with the vehicle during the journey; but the latter are open-sided (Figure 13.1), and drivers are accommodated in a separate carriage with conventional train seating. On the other side of the Channel vehicles are driven off the train for the onward journey. Each train is 745 metres long, can carry up to 32 trucks and 56 drivers and driver's mates, travels at up to 140 kilometres per hour, and transits the tunnel in 35 minutes (Eurotunnel, 2018). Many would argue that this is the easiest way to cross the English Channel.

A longer-distance service is operated across the Swiss Alps. RAlpin operate a rolling highway between Freiburg im Breisgau in Germany and Novara in Italy, a distance of 414 kilometres (or 435 kilometres by road) (Figure 13.2). The journey takes

Figure 13.1 The Eurotunnel shuttle fully loaded with trucks

SOURCE Reproduced by kind permission of Eurotunnel

10 hours, which allows a driver to take statutory rest, and Swiss customs formalities can be avoided. The service is commercially attractive to truck operators as the Swiss Federal Government imposes a tax on HGVs transiting the country, based on emission level, tonnage and mileage. It is reasonable to assume that one of the aims of this tax is to encourage trucks to use the RAlpin service (which was commissioned by the Swiss Government) and in this they have been successful, taking over 100,000 vehicle journeys off of the road in 2017 (RAlpin, 2018).

Bulk and break-bulk haulage

Bulk haulage is an activity in which the heavy carrying capacity of rail gives it a very clear economic advantage over roadfreight, and it can be cheaper over a distance as short as 15 kilometres.

Bulk commodities such as aggregates are hauled in full trains of open wagons by one or more locomotives. Loading and unloading can be by crane using a scoop or grab, or a train may be loaded by a purpose-designed system of chutes into a hopper wagon and discharged by gravity on arrival. An example of the latter is the

Figure 13.2 Trucks being loaded onto the RAlpin rolling highway for transit across the Swiss Alps. An empty train can be seen in the foreground and fully loaded train ready for departure behind

SOURCE Reproduced by kind permission of RAlpin AG

Merry-Go-Round service for coal, in which wagons are loaded from an overhead bunker (formerly at mines but now in a port) and unloaded by driving the train at a constant 0.8 kilometres per hour over an unloading pit, the bottom doors being opened and closed by cams as the wagon reaches the start and end of the pit. The coal is then removed from the pit by conveyor. Unloading takes about 30 minutes, and if all is well there is no need for the train to stop (Glover, 2013).

Cars are an example of break-bulk freight. They are driven onto trains from the rear via purpose-built loading ramps.

Other commodities carried in this way include:

- on open wagons:
  - minerals, eg iron ore;
  - biomass;
  - scrap metal;

- on flat wagons:
  - construction materials;
  - timber;
- in covered wagons:
  - potash;
  - domestic waste;
- in tank wagons (Figure 13.3):
  - oil and petroleum;
  - chemicals;
- on specialized wagons:
  - cars;
  - nuclear waste.

Whilst, as always, generalizations should be treated with caution, the minimum weight likely to be needed to make a bulk haulage train viable is about 700–800 tonnes. Most such trains in the UK are in the region of 1,500 to 2,500 tonnes, but up to 5,000 tonnes is feasible. The heaviest trains in Europe are iron ore trains operating between Sweden and Norway, of approximately 8,500 tonnes.

Figure 13.3 A bulk haulage train, in this case of tank wagons, passes through a rural station

SOURCE Reproduced by kind permission of Freightliner

Intermodal

This is the other main activity for which railfreight is likely to be economically viable, and for which it is used on a large scale in the UK, representing over a third of railfreight traffic. The rail element of intermodal traffic in Europe has grown steadily in recent decades, by 145 per cent between 1990 and 2014 (Monios and Bergqvist, 2017). It can be defined as a journey during which a unit load, such as a shipping container, is transported by at least two different modes. A common example would be a container arriving from the Far East by ship, which is then moved to an inland terminal by rail, and finally delivered to its destination by road.

Dr Richard Beeching, former Chairman of British Railways, has been much criticized for wielding the 'Beeching Axe' of the 1960s, which resulted in the closure of about a third of Britain's railways and over half of its passenger stations. One thing he got right was predicting a future for rail transport of containers. Freightliner ran its first train in 1965, became a separate company from British Rail in 1968, and is still in business today as one of Britain's largest railfreight companies.

Figure 13.4 Intermodal transport – a train of 40-foot containers

SOURCE Reproduced by kind permission of Freightliner

Most intermodal movements in the UK are indeed of containers: 80 per cent of Freightliner's UK intermodal business is on behalf of shipping lines. They operate from the ports of Felixstowe, Southampton, London Gateway and Teesside to a range of inland terminals from Coatbridge to Cardiff, and most deliveries are made to within about 30 miles of the nearest terminal. Road transport can be provided by Freightliner, or arranged independently by the customer. Container storage is offered as an option at most terminals, and return of the empty container to the port or elsewhere (known as restitution) is part of the service provided.

A next-working-day service for delivery of a container from port to destination is available across most of the country, which will be fast enough for almost all practical purposes. Costs are by commercial necessity competitive with road transport.

Longer-distance intermodal trains operate internationally, a very long-distance example being that between China and Europe. The service connects 35 cities in China with 34 centres in Europe, including Barking (East London) via the Channel Tunnel. About 90 per cent of containers on the 11,000 kilometre route travel via the Trans-Siberian Railway, the remainder via Kazakhstan (International Rail Journal, 2018a). Transit times are 14–17 days, which is significantly quicker than by sea, whilst offering a much cheaper option than airfreight for larger consignments.

Figure 13.5 An intermodal terminal in operation. On the left of the picture can be seen rail wagons, both empty and loaded with containers. To the right are stacked containers, and some loaded on road vehicles for movement to or from the terminal

SOURCE Reproduced by kind permission of Freightliner

Swap bodies

A variation for intermodal transportation is to use a swap body rather than a container. This resembles a conventional truck body, except that it is designed to be lifted off the trailer chassis and onto a rail wagon, or vice versa. They are usually curtainsided, which permits side loading and unloading, and subject to the railway loading gauge they can be larger than shipping containers – critically in the UK they can be built with a 2.4 metre door opening, permitting transport of two standard 1,200 millimetre pallets side by side. They also have a lighter tare (or empty) weight. Their disadvantages include higher initial and maintenance costs, and that they are not stackable so cannot be used for seafreight.

As a result, they tend to be preferred for intermodal movements involving only road and rail transport; containers are preferred where seafreight is also involved.

In the UK, several retailers use swap bodies for transport between the Midlands and Scotland, often backloading with goods from Scottish suppliers. Many of these services operate from the Daventry International Rail Freight Terminal (DIRFT), a purpose-built intermodal terminal (Figure 13.6). The economic reality is that services operating from such a terminal are economically viable, whereas a journey that required a roadfreight movement of a swap body trailer both before and after the rail leg, including two lifting operations, probably would not be.

Figure 13.6 The Daventry International Rail Freight Terminal. Swap bodies can be seen on the right loaded onto road trailers, in the centre being hauled up a ramp to the railhead, and loaded onto rail wagons on the left. A swap body in the process of being loaded can be seen to the right of the trains

SOURCE Reproduced by kind permission of Prologis

The economics for longer-distance movements are, however, very different. For example, Transmec offer a daily swap body service between the UK and Italy, with a 48-hour transit time in either direction, prices for which compare well with road-freight.

Piggyback system

This is an intermodal system in which complete semi-trailers from artics are loaded aboard wagons for transportation. Technically, the rolling highway is a piggyback system, but in this context trailers are carried unaccompanied by driver or by tractor unit.

The main advantage over a swap body system is that specialized equipment is not required – conventional trailers can be used. However, there is a significant cost penalty due to the weight, and space occupied by, the chassis and wheels of the trailer. The system is therefore rarely used in Europe, and is not feasible at all in the UK due to restrictions of the loading gauge.

Single wagon and part train loads

The traditional model for railfreight, with single wagon or even part train movements, is in long-term decline. This model operated via marshalling yards (also known as switching terminals), with trains comprising wagons for multiple customers being broken down and reassembled for onward movement. A study (European Commission, 2015) found a decrease of 15–20 per cent since a previous study five years earlier, citing factors including poor reliability, poor profitability for operators, the closure of marshalling yards (30–40 per cent in the previous decade) and the closure of private sidings. It was concluded that there is now a vicious circle, with reduced infrastructure causing a reduction in traffic, which in turn causes further reductions in infrastructure. In Germany such services still account for 39 per cent of rail traffic, and in some countries such as Austria intermodal wagons form part of these trains (Monios and Bergqvist, 2017), which no doubt will prolong their use, but in the UK they have almost disappeared. Single wagon services still operate on a large scale in the USA and Canada: the Belt Railway Switching Terminal in Chicago handles an average of 8,400 rail cars, equivalent to 65–80 kilometres of train, every 24 hours (Belt Railway, 2019).

My overall advice is that if you are considering a switch to rail in Europe, this is a business model of the past rather than of the future, and is unlikely to be the best long-term solution to your needs. In North America, however, it continues to be a viable alternative.

Freight on passenger trains

This is another service model that has suffered a long-term decline. The option for station platform staff and guards of passenger trains to handle parcels and other items into and out of the brake van at stations is no longer feasible – the delay to the train would be unacceptable to passengers, and staffing levels do not permit this: indeed in some cases there may be neither a guard nor any platform staff present.

There are a few well-publicized examples that have been set up in recent years, such as the transport of live lobsters in special cold crates from Penzance to London Paddington to serve upmarket restaurants in the Capital (GWR, 2015). However, these examples are rare. Unless you have a similar very specialized need, you are unlikely to be able to make use of this option.

Fast rail services

An interesting new development is the introduction of a fast railfreight service in Italy in November 2018. Mercitalia, a subsidiary of the state-owned railway company, is now operating a service between Caserta (north of Naples) and Bologna five nights per week. It covers the road distance of 546 kilometres in 3.5 hours. The trains comprise 12 wagons, modified to carry roll containers, and the train's payload is equivalent to 18 articulated trucks (International Railway Journal, 2018b).

Previously, high-speed lines have been the preserve of passenger services such as the French TGV and Japanese Bullet Train. It will be interesting to see whether this service is a commercial success – if so, it could lead to a rapid expansion of such services elsewhere.

Access to railfreight services

In the UK, I have already mentioned Freightliner as a major provider. Logico act as their general commercial division and would be the first point of contact. GB Railfreight and DB Cargo are others, the latter operating many services internationally, including the services to and from China. Intermodal services can be accessed through these providers, or in some cases through a third party – for example the Transmec service to Italy described earlier. It is possible to begin a swap body or container operation without new facilities at your premises, and whilst I would always recommend planning well in advance, the time taken should not be great.

For bulk haulage, however, there are potentially substantial issues that need to be resolved before the service can begin, as discussed below.

Provision of rail terminals for bulk haulage

If you are fortunate, you may have an existing rail connection to your premises, or access to one nearby. Alternatively, in some remote areas such as northern Scotland it may be possible to load trains whilst they stand on the main running line, but opportunities for this solution are understandably rare.

However, if you are not so fortunate, you will need to establish a rail terminal. This will as a minimum require a siding with a connection to an existing running line, and sufficient space for loading/unloading equipment such as cranes, and storage of freight awaiting loading or onward transit.

> Fortunately, Network Rail (the body owned by the UK Government responsible for physical rail infrastructure such as lines, bridges, and signalling) may be able to assist with providing a suitable location. They already have a freight estate of more than 100 sites, on which they own the freehold, but lease the site to another party who has installed the infrastructure for a railfreight terminal. Over 50 further sites are available nationwide (as at 2018) for which similar arrangements can be made. Capital for investment in civil engineering works is very unlikely to be provided by the Government, so they are not able to offer turnkey solutions, but subject to your financial strength it may well be possible to obtain private sources of finance, and they may be able to provide an initial period at reduced rental to assist with cash flow.

If you are interested in such possibilities, I would recommend consulting the Network Rail website in the first instance.

Whether using this option or a site you have identified yourself, you should consult Network Rail at an early stage as to its suitability. New points from the main running line will need to be installed, and probably also a crossover to allow access to tracks in both directions (technically known as 'up' and 'down' lines). Changes will also need to be made to signalling to control access.

It is rare for a freight terminal to be equipped with overhead wires for electric locomotives – they can be a serious safety issue, and would of course render the loading and unloading of wagons from above impossible.

Network Rail will need to ensure that the proposed route between the terminal and expected destinations is suitable for the projected traffic. Amongst other factors, they will need to confirm that the loading gauge is sufficient, and that axle weights will not exceed safe limits for the line. Some routes are already heavily congested, and the introduction of freight trains might be an irresolvable problem. Devising train paths, as they are known, for a variety of passenger trains (eg express, fast commuter and slow commuter), each with different speeds, acceleration and stopping patterns, is a difficult enough exercise, especially where there is only one line in each direction. To incorporate a freight train, with slower speed – typically a maximum of 100 kilometres per hour for a bulk haulage train compared to 200 kilometres per hour for an express passenger train – and slower acceleration but no stops, may not be possible, especially at peak hours, and many freight trains must therefore operate at night.

It will help the process of establishing your terminal if there is an existing rail connection to the site, or the site is at least close to an existing track. The laying of a long stretch of track to provide access may prove prohibitively expensive, especially if this requires the purchase of additional land, and will be impossible in urban areas. The new rail terminal will also need good road access, unless it is located, for example, at a quarry, and of course sufficient space for loading/unloading equipment and storage of goods whilst awaiting handling.

Local opposition

The attitude of the public and some parts of the media towards railways can be somewhat contradictory. If questioned, most people will say that they are completely in favour of transferring freight from road to rail. However, if asked whether particular additional railways or freight terminals should be built, they will probably have the opposite opinion – attitudes toward HS2 and a railfreight terminal close to the M25 on the site of the former Radlett Airfield are examples.

I will not express an opinion on those particular cases, but my own experience in leading the project team to establish what became the first private purpose-built Channel Tunnel Rail Terminal in England will serve as an example. We looked at numerous potential sites, some of which had to be rejected due to high asking prices, which the business case could not support, or operational difficulties we could not overcome. Others met with objections from local press and public opinion, citing reasons such as the creation of insufficient jobs and in one case the idea that we should not redevelop the site of a disused coal mine as it should be re-opened to produce coal. Fortunately, we eventually located a suitable site at Corby, on a brownfield site formerly used by British Steel, where the local council saw the long-term advantages of attracting such a facility to the area, and the terminal proved a great success.

A valuable suggestion would be to find a brownfield site if possible, preferably with an existing rail connection, and to gather as much local support as you can at the earliest possible stage. Proposing to create jobs will also help. Above all, you may need to be persistent, and not be disheartened by early setbacks.

I predict that once the terminal is up and running, almost all opposition will quickly be forgotten, people will recognize the advantages of railfreight, and you will receive a great deal of credit for shifting freight from road to rail!

Government subsidy

In the UK, subsidies are available from the Government to support the transfer of freight from road to rail (or to inland waterways). The scheme is known as the Mode Shift Revenue Support scheme, or MSRS, and will operate until at least 2020.

The aim of the scheme is to negate the additional cost of operating a more environmentally friendly mode, so that if a change to rail would otherwise result in an additional cost relative to roadfreight, the subsidy will ensure that you are not financially penalized for doing so.

The scheme is subdivided, with different mechanisms for intermodal and bulk traffic. For the former, the subsidy is based on a per unit figure, whereby a fixed sum is paid for each container or swap body moved between two locations. Maximum subsidies are specified based on 18 regions of the UK: for example, a container from the Port of Southampton to London could attract a maximum subsidy of £19, to Birmingham £42, but none if shipped to Glasgow. For bulk traffic, the calculation is made on a case-by-case basis, with the difference in price between the two modes being the maximum possible grant payable. In all cases, you will not only need to prove the financial case, but also the environmental benefits of changing modes. The proposal will need approval in advance, but payments are made in arrears to protect the interests of the taxpayer.

In Scotland and Wales, Freight Facilities Grants may be available to assist with capital investment.

The information in this section is based on *The Guide to Mode Shift Revenue Support (MSRS) Scheme* (Department for Transport, 2015).

Inland waterways

Inland waterway transport (IWT) is probably the oldest mode of freight transport, the earliest example probably being far into pre-history when a hunter-gatherer used a rudimentary canoe to bring home the product of a successful hunt. In the Middle

Ages, rivers were again the primary route, the roads of the time being unsuitable, and the canals had their heyday in the early period of the Industrial Revolution before the advent of the railways.

One way in which waterways differ from roads and railways is that the latter are used only for transport of freight and passengers. Waterways, however, have other functions, including as sources of water for drinking, crop irrigation, and industry; hydro-electric power generation; drainage; and recreation, such as swimming, fishing and sailing. The needs of these other uses may conflict with the needs for use as transport, either permanently (eg by building of dams) or temporarily by creating a shortage of water, so that vessels are subject to grounding; or by creating an excess of water, so that vessels cannot pass below bridges. These factors should be taken into account if considering the use of IWT.

IWT in the UK

> If you are considering using the 18th century narrow canals of the UK for long-distance transport, I have one simple piece of practical advice – forget it!

The canals of Britain had been in long-term decline for many years, but the critical blow occurred in the severe winter of 1962–63, when the canals froze solid for several months. Many customers switched modes from necessity and did not return. With the growth of the motorway network they became unable to compete. As an example, a round-trip from London to Stoke on Trent can be completed in a day by an artic if timed to avoid congestion; by canal the round trip would take approximately 185 hours based on a calculation by Canal Plan (2018) – about a month assuming nine hours per day, five days per week. The cost of paying the bargee for a month would make this completely unviable.

There are some examples of the use of more major waterways for IWT in the UK. They include:

- 700,000 tonnes of domestic refuse are transported annually by Cory Environmental's fleet of seven tugs and 47 barges on the Thames, to landfill or waste processing sites (Freight by Water, 2018).
- Sand and gravel aggregates from a quarry at Upton on Severn are shipped by barge on that river and the adjoining Sharpness Canal, each barge being able to carry at least 350 tonnes (Severn Boating, 2018).
- Port Salford, a container terminal located on the Manchester Ship Canal, handled 30,000 containers in 2014 (Robins, 2015).

The MSRS scheme (see above) is available for IWT as well as railfreight, and if you are able to identify an opportunity to utilize the waterways this will have clear environmental benefits. These opportunities will, however, be rare, and for many readers of this book the option of IWT will not be in any way realistic.

IWT in continental Europe and elsewhere

Outside the UK, the situation may be very different. IWT has a modal share of about 6 per cent in Europe, which was roughly constant between 2001 and 2012 (European Court of Auditors, 2015).

Estimates of the total length of navigable waterways vary, in part due to differing definitions of 'navigable'. Wiegmans and Konings (2017) estimate 600,000 kilometres worldwide, including 52,000 kilometres in Europe, of which 8,000 kilometres are relevant for freight transport. The backbone of the European system is the River Danube flowing through Germany, Austria, Hungary, Serbia and Romania, which is 2,850 kilometres; and the Rhine, 1,300 kilometres in Switzerland, Germany and the Netherlands. The former can be unreliable due to varying water levels and serves a less industrialized area than the Rhine. Together with its tributaries and adjoining canals, the Rhine axis as it is known carries about 70 per cent of European IWT.

Other important waterways include the Seine and Rhone in France, and the extensive wide canal systems in Holland, Belgium and some parts of France and Germany. The Rhine-Scheldt canal linking the major ports of Rotterdam and Antwerp is one of the busiest waterways.

IWT tends to be most competitive for carriage of similar types of freight to rail, ie intermodal and bulk – the latter including tankers (Figure 13.7). A Rhine vessel can be up to 110 metres long, 11.4 metres wide, and have a draught (ie the distance between waterline and the bottom of the boat) of 3.6 metres. This permits a load of 3,500 tonnes or 208 20-foot containers.

Two main types of vessel are in use. They may be either self-propelled, or consist of a number of so-called dumb barges propelled by a pusher vessel (Figure 13.8). The latter offer an economy of scale: a pusher convoy to Duisburg can carry 18,000 tonnes of coal (Wiegmans and Konings, 2017). However, they are more difficult to operate due to the need to marshal the barges for loading and unloading. Monios and Bergqvist (2017) provide the following figures for the number of vessels in Western Europe:

Figure 13.7 The 2,750 tonne Dutch tanker barge *Zuidwal* navigating a canal

SOURCE Reproduced by kind permission of Mercurius Group

Self-propelled	Dry cargo	6,753
	Tanker	1,999
	Total	8,745
Dumb barges	Dry cargo	3,117
	Tanker	155
	Total	3,272
Pushers	1,039	

Much of the container traffic relates to the so-called extended gate concept. Containers are transhipped in the great ports of Rotterdam and Europort, for onward shipment to an inland terminal, customs and other formalities only being conducted on arrival. These terminals can be as far inland as Nuremburg and Liege, and some are very large – Tilburg has a capacity of 300,000 20-foot container equivalents per year, which is about a third of the size of the Port of Rotterdam (Monios and Bergqvist, 2017)

Figure 13.8 Pusher vessel and barges on the Danube

SOURCE Pixabay

Figure 13.9 The Dutch inland container ship *Lahringen*. The vessel can carry 208 20-foot containers

SOURCE Reproduced by kind permission of Mercurius Group

Much of the bulk freight is also transhipped in the major seaports, such as coal for power stations. In the case of tankers, several refineries and chemical plants are located close to ports, and it is a refined or manufactured product that is transported by IWT. Palletized goods are not commonly transported by water, but recently there have been attempts to utilize the Belgian canal network for transport of building materials, which is demonstrating clear potential (Monios and Bergqvist, 2017).

In Europe most barges are operated by small businesses, each owning just one to three vessels (this contrasts with the US, where just five major players dominate the market). There is over-capacity in the market, which results in low prices and therefore low profits. Another problem is recruitment, with a shortage of personnel and an ageing work-force. Automated or semi-automated vessels may provide a long-term solution.

IWT can be economic over short distances, as is demonstrated by the transhipment of containers in Rotterdam, more than half of which are moved only to other locations in Holland rather than internationally. However, by its nature it is better suited to longer-distance transport, as illustrated by the figures in Table 13.1, based on data provided by Wiegmans and Konings (2017).

It can be seen that there is a clear trade-off: the reduction in cost will more than compensate for the increase in transit time for material that is not time-critical. There is also a clear advantage for some very large items, such as blades for wind turbines, which can be carried on a barge but not on a road vehicle or rail wagon. IWT is, however, much less well suited to urgent and just-in-time deliveries.

Outside Europe, IWT plays a major role in some countries. Examples include China, where an estimated 165,000 vessels ply the trade (Monios and Bergqvist, 2017); Russia; Brazil, where the Amazon and its tributaries are the only means of access to some areas; and the USA, especially on the Great Lakes and Mississippi and Ohio rivers. In these countries, and in some parts of Europe, inland waterway transport is a very real option for the right type of cargo. If your business is located by a suitable waterway, it should seriously be considered as an option.

Urban delivery by canal

The use of canals in Venice by all professions, from midwives to undertakers, is well known, but this is driven by the unique historic features of the city. There are also electrically powered boats that make deliveries of beer in Amsterdam and Utrecht; and even boats selling ice cream on the Norfolk Broads. However, very special circumstances apply in all these cases, and opportunities to replicate the model elsewhere will be extremely rare.

Table 13.1 Relative costs of road, rail and inland waterway transport

		Saving/(on-cost) relative to road			
		Road	**Rail**	**Waterway**	**Distance by road (km)**
Rotterdam to Heidelberg	Containers	N/A	(5%)	27%	515
Transit time (days)		1.5	2.5	3.5	
Rotterdam to Vienna	Bulk liquid	N/A	20%	33%	1,157
Transit time (days)		2	4	8	
Austria to Romania	Cars	N/A	7%	65%	1,070*
Transit time (days)		3	6	7	

SOURCE Savings and transit times from Wiegmans and Konings (2017). Distances calculated by author
* Assuming Vienna to Bucharest

Fixed installations

An often-forgotten option for inland transportation is the fixed installation, which makes a more substantial contribution than is usually realized. In the UK, for example, 140 million tonnes of material were moved by pipeline in 2012, more than by rail or by water (Department for Transport, 2017).

Pipelines are used for liquids and gasses, and some very long pipelines carry these products, for example from Russia to Western Europe. In the UK there are pipelines linking major oil refineries with inland depots, and dedicated pipelines for aviation fuel into Heathrow and Gatwick. Pipelines have also been used for transport of coal, by mixing coal dust with water to form a slurry.

Conveyors are used mainly to support the agricultural, mining and quarrying industries, moving ores, aggregates, coal, farm produce and other commodities (Figure 13.10).

Figure 13.10 Outdoor conveyor used for transport of coal

SOURCE Pixabay

Each of these obviously has a very high investment cost, to which potential costs for decommissioning must be added. However, once built the running costs of pumping and maintenance, measured on a per tonne basis, will be very low, and their use is much easier than making arrangements for trucks, trains or barges.

Pipelines and conveyors do not of course need to be of great length, and may be useful for moving goods between sites, or to or from a transport terminal. Once again, there will be few cases in which this represents the best option, but you should be aware of the possibility if you are moving very high volumes of material continuously on the same route.

Conclusion

As I explained earlier, for many types of inland freight movement, such as the delivery of a single parcel or pallet, there is only one sensible choice – the use of roadfreight. However, for bulk haulage there can be very clear financial and environmental advantages in looking at the alternatives of railfreight and, in some geographical areas, inland waterways. Intermodal transport, using either (or indeed both) modes, is growing in popularity, and will no doubt continue to do so. Pipelines and conveyors also make a major but often forgotten contribution. The next chapter will look at the main choices for overseas transportation – seafreight and airfreight.

14 Seafreight and airfreight

In previous chapters I have discussed the options for inland transport. These have ranged from local take-away food deliveries using a pedal cycle or small van, to bulk haulage of several thousand tonnes of aggregates by train or barge. Road and rail services do, of course, operate internationally, the intermodal rail service between China and Europe probably being the longest route.

However, for most long-distance, and especially inter-continental, movements the options are likely to be restricted to seafreight and airfreight. I will talk about the latter in the second part of the chapter.

The range of ship types is probably even more extensive than the range of road vehicle types described in Chapter 8. However, most readers of this book will not need to concern themselves with large tankers or bulk carriers, so I will mainly restrict myself to those which will be encountered most frequently: roll-on, roll-off; containers; and general cargo.

Roll-on/roll-off

Roll-on/roll-off (ro-ro) means that freight is rolled onto and off a ship on wheels. Most readers will have travelled by ferry, perhaps from Dover to Calais, and the ferry is an example of a ro-ro ship. These operate on short sea crossings throughout the world and are often the only means of access to islands. In other cases, they act as a short-cut, where the land journey would be much longer. Examples include the Gourock to Dunoon ferry across the Clyde Estuary in Scotland, a 20 minute crossing that saves a 132 kilometre road journey; and the ferry from Brindisi in Italy to Igoumenitsa in Greece, a 7.5 hour crossing that saves a road journey of no less than 2,600 kilometres.

In some countries ro-ro ferries are integrated into the road system. An example is Norway, where travelling the length of the country using the E30 main route involves seven ferry crossings, such as that across the Sognefjord between Ytre Oppedal and Lavik – a very pleasant journey on a clear day but also of vital importance to the area. They are also of vital importance in island nations such as Indonesia.

Most ferries can accommodate HGVs, including artics. There are a few exceptions, such as the Ballyhack to Passage East service in Southern Ireland, which accepts vehicles to a maximum of 3 tonnes (Passage East Ferry Company, 2018). It is therefore wise to check in advance, but problems will be rare.

On shorter routes, artics will usually be shipped as a complete combination, the driver using the ferry as part of their journey. This is known as driver-accompanied shipping. However, it is possible to arrange shipment of unaccompanied trailers, the shipping line being responsible for loading and unloading the trailer, probably using a tugmaster (Figure 14.1).

Many ro-ro services carry both cars and freight vehicles: often the service would not survive commercially without both revenue streams. However, some ferries are designed for freight only (Figure 14.2), and do not accept passenger traffic, except perhaps for a few drivers. Often there will be a choice to be made, between a longer passage on a freight ferry or a shorter crossing on a combined freight and passenger ferry. For example, a journey from Birmingham to Cologne could be made on a driver-accompanied basis via the Channel Tunnel or a Dover to Calais ferry; or on a freight-only ferry, accompanied or unaccompanied, on the Cobelfret Purfleet to Rotterdam route. The latter is a 12 to 16 hour crossing, with only four crossings per

Figure 14.1 An unaccompanied trailer being offloaded by a tugmaster from the ro-ro ship *Yasmine*

SOURCE Reproduced by kind permission of CLdN Group

day, and will be more expensive. However, it will save about 150 miles of road mileage, the cost of which would exceed the additional ferry cost. If separate arrangements can be made for haulage on the two sides of the North Sea so that the trailer is shipped unaccompanied and the artic unit is not tied up during the voyage, the savings will be even greater. There is, therefore, a classic case of a trade-off between cost and speed.

Hazardous goods are more likely to be accepted on a freight-only ferry, but rules vary between routes according to factors such as the expected sea conditions. It should also be remembered that a ship's captain's first duty is to the safety of his crew, and therefore they have an absolute right to reject any cargo on any voyage. I recall this happening with a full load of car batteries, normally accepted on the particular service, but refused on one particular occasion when severe gales were forecast.

Livestock movements are less likely to be accepted on longer routes.

Freight such as caravans, agricultural equipment or cars can of course be shipped as ro-ro cargo. Some services therefore carry a mix of ro-ro, container and general cargo, for example the Atlantic Ro-Ro Carriers service between Russia and the USA (ARRC, 2018). Large quantities of cars, such as imports to Europe from Japan, are carried on specialist car carriers, but a shipment of a few new cars to a dealership in the Hebrides would be accommodated on a shared-user ro-ro ship.

Figure 14.2 The freight-only ferry *Celine*. Built in 2017, with a gross tonnage of over 74,000 tons, it has capacity of almost 8,000 loading metres. This means that if all vehicles loaded aboard were arranged end to end, the line would stretch for nearly 8 kilometres

SOURCE Reproduced by kind permission of CLdN Group

Container shipping

If your company is importing or exporting its products, it is likely that you will be using containers. Whilst a form of containerization can be traced back to the 18th century, modern usage started in the Second World War in the US Army.

> In 1956 US trucking magnate Malcom (*sic*) McLean devised the standard metal container. He found that loading a container ship cost $0.16 per tonne, compared to $5.83 per tonne for loose cargo. Ships also spent much less unproductive time in port, and theft of freight reduced dramatically (Economist, 2013).

Usage of containers (sometimes known simply as boxes) has grown enormously, the biggest trade route being Asia to North America, with over 23 million TEU (see below) in 2013. Major users include Walmart, with approximately 825,000 TEU into the US alone (from all sources) in 2016; and the Dole food company, importing 227,000 TEU of bananas into the US in the same year (Neise, 2018).

The containers themselves

> Two types of container dominate the market – the 20 foot (6.098 metres) and the 40-foot (12.192 metres). The former is the basis for measurement of quantities, which are expressed in TEU, ie twenty foot equivalent units. Thus, a ship able to carry 1,000 40-foot containers will be described as having a capacity of 2,000 TEU.

It should be noted that imperial (rather than metric) measurements are the world-wide standard for containers. In recent years there has been a move towards higher cube containers, with a height of 9 feet 6 inches (2.896 metres) rather than 8 feet 6 inches (2.591 metres), and/or a length of 45 feet (13.716 metres). There is a potential problem with the movement of the latter on UK roads, but it is possible to use a Euro-casting designed by the Geest company which ensures that the vehicle and trailer remain within maximum permitted dimensions (Lowe and Pidgeon, 2018).

Ten foot boxes were in common use in the former Soviet Union but are rarely seen today in the west; 30-foot containers are also rare.

The maximum weight capacity of containers varies slightly according to their construction, but typical figures are 28.2 tonnes for a 20-foot and 26.6 tonnes for a 40-foot (Neise, 2018). For heavy freight, such as tinned fruit, the 20 foot box is therefore the best choice; for lighter items such as plastics a 40-foot or high cube would be the better option.

There are also specialized containers. Reefers or refrigerated containers are used for perishable goods, drawing power from electrical points aboard the ship. Tank containers are used for liquids, such as chemicals. Some non-hazardous liquids, such as wine being shipped in bulk for bottling on arrival in Europe, may be shipped in strong bags (called flexitanks) which occupy most of the inside of a conventional container, in some ways reminiscent of a very large wine box. Specialized containers are used for products such as hanging garments.

Container loading

It is the responsibility of the shipper to load a container to avoid damage to the goods. It is worth remembering that conditions encountered during a sea passage will be much more demanding than those on a motorway. It is therefore recommended that a minimum of double wall cartons is used for all container shipments by sea (see Chapter 7). Within those cartons, fragile items should be wrapped in a minimum of 5 centimetres of bubble wrap.

You should always plan stowage (also known as container 'stuffing') before starting the process. The following are useful guidelines:

- Examine the container for damage and cleanliness before loading.
- Packages should be neatly stacked, with the heaviest boxes on the bottom.
- Ensure that all heavy goods are not loaded at one end. Goods should be evenly spread across the floor of the container if possible.
    - If not, eg in the case of a single item of machinery, it should be placed centrally and secured with straps, nets, inflatable bags, and/or wooden material known as dunnage.
    - If there is a gap between stacks of goods, the stack should be positioned centrally, either along the length or across the width of the container.
- Any sharp corners should be covered to ensure they do not damage other items.
- Any noxious goods should be sealed.
- Do no load odorous goods in the same container as vulnerable goods – onions and high-fashion clothing do not mix well.

On completion of loading, the container should be sealed. I strongly recommend the use of a steel bolt seal (Figure 14.3). Whilst plastic or wire seals are cheaper, a high-security

Figure 14.3 A steel bolt type high-security container seal

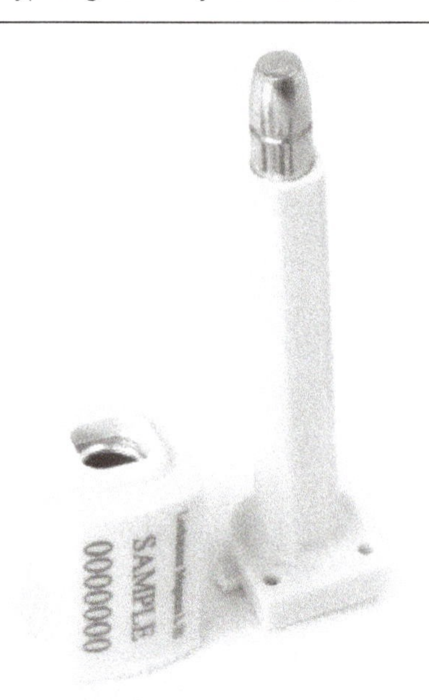

SOURCE Reproduced by kind permission of Universeal

seal will cost only a few pounds and be well worth the money. The seal number should be recorded on the relevant paperwork, and many people also take a timed and dated photograph.

On receiving a container, be careful to check the seal number against the paperwork before the seal is broken.

> Before receiving a container, ensure that you have a good set of long-handled bolt cutters. It is not feasible to cut through 8 millimetres of low-carbon steel with pliers, and difficult and time-consuming with a miniature hacksaw – especially if you have to drive to the local DIY store to buy one!

Verified gross mass of containers

> A major factor in the shipwreck of the MSC *Napoli* in 2017 was the under-declaration of the weights of containers. The crew of 26 were forced to abandon ship and take to lifeboats in force 11 winds and 40-foot waves, and were extremely lucky to survive. The salvage operation cost £120 million (BBC, 2017). Safety at sea is not a matter to be trifled with, and great care should be taken to ensure that all regulations are adhered to rigidly.

Since 2016, the International Convention for the Safety of Life at Sea (SOLAS) has made it mandatory to declare a verified gross mass (VGM) for every container, so that the captain can arrange stowage in a manner that does not adversely affect the ship's stability. This can be determined by weighing the complete loaded container, or by weighing each item of cargo before it is loaded, and adding the total to the tare (empty) weight written on the container door. It is not permitted to estimate weights. There is one exception, whereby 'Individual, original sealed packages that have the accurate mass of the packages and cargo items (including any other material such as packing material and refrigerants inside the packages) clearly and permanently marked on their surfaces, do not need to be weighed again when they are packed into the container' (World Shipping Council, 2015). It may, however, still be prudent to check weigh a sample of the items.

It is the responsibility of the shipper to provide the VGM, although a representative may be authorized to do so on their behalf, such as a 3PL. Ships are not permitted to load containers without a VGM.

Vessel routing

Container shipping is dominated by major lines: APM-Maersk, MSC and CMA CGM control 41 per cent of the world's capacity between them (Neise, 2018). Ships on deep-sea (or intercontinental) voyages tend to be routed so that containers are collected from a number of ports, and discharged at a number of ports, at either end of the voyage. Services operating on fixed routes, whether long or short, are known as liner services. For example, Maersk (2018) has routes including:

- Ningbo, Shanghai, Xiamen, Hong Kong, Tanjung Pelepas (Malaysia), Colombo, Felixstowe, Rotterdam, Bremerhaven;
- St Petersburg, Bremerhaven, Rotterdam, Antwerp, Cartagena (Colombia), Manzanillo (Panama), Guayaquil (Ecuador), Balboa (Panama);
- Tanjung Pelepas (Malaysia), Singapore, Brisbane, Tauranga (NZ), Lyttleton (NZ), Port Chalmers (NZ), Tanjung Pelepas.

Some deep-sea routes are operated by very large container ships, some with a capacity of over 20,000 TEU.

There are also what are known as feeder routes, operated by smaller ships, which transport containers between major ports on the deep-sea routes and smaller ports. Maersk, for example, have feeder routes between Bremerhaven and Norwegian ports. Others are operated by feeder specialists, such as Unifeeder, whose network includes ports such as South Shields in the UK and Rauma in Finland that are unlikely to be served by deep-sea vessels.

Figure 14.4 Large container ships in the Port of Southampton with extensive container park and rail terminal in foreground

SOURCE Reproduced by kind permission of Freightliner

Figure 14.5 Feeder ship *Elbstrom* being unloaded in the Polish port of Gdynia

SOURCE Reproduced by kind permission of Unifeeder and Dawid Bujewski. Picture by Dawid Bujewski

Short sea routes, as the name implies, are operated by container ships over shorter distances; some of these also act as feeder services. They offer an alternative to road and rail transport. For example, if you are importing tinned tomatoes from Southern Italy to the UK, a container service from a port such as Salerno will probably be the cheapest routing, although transit times will be greater. As so often, there is a choice between cost and speed.

Full container load and less than container load traffic

Most container movements consist of a full container load consigned from A to B – this is known as FCL. However, you may have insufficient freight to fill a container – maybe there is only one tonne. This is known as less than container load (LCL). In this case it is best to engage the services of a freight forwarder, who can arrange for the goods to be combined with those of other shippers to make up an FCL. Charges will be calculated according to the weight of the consignment, or its volume, typically on the basis of a cubic metre equating to one tonne, whichever is the higher. At the port of arrival, the container will be de-stuffed to permit delivery of each consignment.

> In some ports in the world, goods are not handled as carefully as you might wish. If using LCL services, it is wise to ensure that your items are very well packed.

Handling in ports tends to be more expensive than elsewhere, and LCL involves at least two transhipment movements. Booking a 20 foot box may be cheaper than an LCL shipment, so might well be the most economical option even if the container is not full, and has additional advantages in reduced risk of theft or damage. As a broad generalization, this is likely to apply to consignments of more than about 15 tonnes or 15 cubic metres.

At the smaller end of the scale, LCL shipments are subject to minimum charges, often of 1 cubic metre or one tonne. If you wish to ship, say, 25 kilogrammes, airfreight will probably be cheaper than seafreight.

Container movements to and from ports

There are three main choices for the transporting containers to and from ports – road, rail or barge. Relative numbers vary according to local geography and transport infrastructure. Some examples are given in Table 14.1.

The easiest option is to engage the shipping line to make the movement. Each will have a matrix of what are called grid rates, which state the delivery costs of containers to different areas. The choice of transport method is then a decision for the shipping line.

Table 14.1 Transport mode for container movements from port

Port	Road	Rail	Barge
Felixstowe	73%	27%	0%
Shanghai	92%	0%	8%
Hamburg	55%	43%	2%
Antwerp	58%	7%	35%
Barcelona	90%	10%	0%

SOURCE Felixtowe: Port of Felixtowe (2018); other ports: Neise (2018)
NOTE Excludes onward movement by sea

Alternatively, you can use what is known as merchant haulage, and engage the services of a specialist road haulier and/or intermodal rail or barge operator (Figure 14.6). If you do so, you will be charged a lift-on/lift-off (lo-lo) charge by the shipping line for loading the container onto the trailer, rail wagon or barge. By commercial necessity, merchant haulage rates tend to be set at a rate which, when added to the lo-lo charge, is a little cheaper than the grid rate. Both grid rates and merchant haulage rates should include restitution, ie taking the empty container back to the port or elsewhere after unloading. I have never personally known a problem to occur with this, but it may be worth asking the question.

It is in theory possible to arrange to collect your own containers from the port, but few companies choose to do so.

Buying seafreight

There are three types of organization from which you can buy seafreight:

- shipping lines;
- non vessel owning common (or container) carriers (NVOCCs);
- freight forwarders.

The first of these speaks for itself – a shipping line operates the ship, and will sell space on that ship to anyone prepared to pay the price they ask. NVOCCs do not themselves own ships, but buy slots on those ships in bulk, which they then sell on to shippers. Freight forwarders also buy and sell freight services, but will also offer a range of other services, such as handling documentation and arranging customs clearance. Confusingly, some companies offer both NVOCC and freight forwarding services.

Figure 14.6 Example of merchant haulage – a container leaving Felixstowe by road for inland delivery

SOURCE Reproduced by kind permission of Goldstar

If you are inexperienced in seafreight, I would recommend that you use a freight forwarder. They know the practicalities of dealing with shipping lines in particular ports, and if they have built personal relationships this will enable day-to-day activities to run more smoothly. They have greater purchasing power than individual shippers, and a small user of container services will probably obtain a cheaper price from a freight forwarder than directly from the shipping line.

The British International Freight Association (BIFA) has about 1,500 members, and you can use their website to find a freight forwarder in your area. The largest worldwide are Kuhne and Nagel, DHL and Sinotrans (Neise, 2018). Some specialize in particular markets, for example FFG Hillebrand for wines, beers and spirits; and some are very small – in extreme cases with just a single employee.

Even large users such as major retailers use freight forwarders for some imports, finding they obtain better rates by doing so.

Rates can be very volatile: an extreme example occurred in 2016 when Hanjin Line suffered a financial collapse, and container rates on routes from Asia increased by 42 per cent the following day (Neise, 2018). If you can fix rates for a period in advance this will probably be a good idea. Rates have been low for some years. There have been numerous mergers and acquisitions as a result, and some lines have reduced the sailing speed of their ships to save fuel.

Surcharges

One feature of the industry is that basic rates may be subject to numerous surcharges, which can amount to hundreds of dollars per container. Before agreeing to use any carrier, it is wise to establish what additional charges will apply, and if and when they may increase. Examples include:

- bunker adjustment factor, to compensate the shipping line for increases in the price of oil fuel;
- emergency fuel adjustment factor, a short-notice version of bunker adjustment factor;
- low sulphur surcharge, reflecting the additional cost of low sulphur emission fuel;
- currency adjustment factor, to account for exchange rate fluctuations;
- peak season surcharge;
- security surcharges;
- war risk surcharges;
- winter surcharges;
- port congestion surcharges.

You may also have to pay terminal handling charges for handling the container within the port.

These charges are a feature of the industry, and do not reflect on any single organization. If you are able to agree an all-in rate to include these charges, then this is clearly the best option. If not, then it is important to be aware of them, and take them fully into account when comparing rates from different sources.

Ensure that you are aware of charges due to any delay you may cause. You will be allowed a certain amount of free time between the unloading of a container from the ship, to the discharge of that container at its destination. If you exceed this, you will be charged:

- demurrage, for additional time spent by the container in the port;
- quay rent, for storage in the port, in addition to demurrage;
- detention, for additional time spent outside the port.

These charges can rapidly escalate if you are not careful, and it is therefore in your interest to unload containers promptly – they are a very expensive way to warehouse goods. Remember to consider possible delays due to customs inspections, the need to pay charges before containers are released to you, and any closures of your premises. A container that arrives in port before Christmas and is unloaded in January, or one that arrives on the first day of the finance director's holiday and is delayed two weeks because no one else is authorized to make a bank transfer for charges, will attract demurrage and quay rent. The secret is in planning ahead: be aware of the estimated arrival time of containers in port, book merchant haulage in good time, be ready to pay charges quickly, and allocate slot times at your warehouse for unloading.

Bill of lading

This is one of the key documents in the shipping industry – the history of similar documents can be traced back to Roman times. In its simplest form, the shipper hands over freight to the shipping line for loading on a ship, who in return provides a piece of paper with all the relevant details. This is called a bill of lading, abbreviated to B/L or BoL. The shipper then posts this to the consignee, who hands over the B/L to the shipping line on the arrival of the ship, and receives the goods in exchange. It is useful in practice to appoint a freight forwarder or other agent in port to do this on your behalf.

If an NVOCC is involved, the shipping line will issue a master B/L to them, and they will in turn issue a house B/L to the shipper in their own name. There are also express B/Ls, which give full instructions to deliver the consignment to a particular address, and no physical documents need to be handed over. Electronic versions of the process are understandably growing in popularity.

Finally, you may see the abbreviation STC on a B/L. This stands for 'said to contain' – if a container is delivered with seal intact the shipping line has no way of knowing the contents, and will not accept liability if the contents are not as described.

General cargo

General cargo is somewhat difficult to define: indeed, definitions do vary. I have described ro-ro and container freight, and mentioned tankers and bulk carriers. There are also other types of specialized vessel. General cargo is in simple terms anything that does not fit into those categories. Examples include steel, timber, machinery or even a yacht. Project cargo is used to describe large items too big to fit into a container, for example a tall radio mast. Such items are often shipped as deck cargo.

There are far fewer liner services for general cargo than in the days before containerization, but they do still operate (Figure 14.7). Examples include those operated by Scotline. Their main focus is on forest products (timber and its derivatives), but they do accept other cargo. Their fleet comprises 14 ships, with capacity from 1,300 to 4,800 tonnes. Regular routes operate from Varberg (Sweden) to a multitude of UK and Irish ports; Latvia to Rochester and Inverness; Wismar (German Baltic Coast) to Rochester; and from Brake (near Bremen) to

Figure 14.7 A general cargo liner service. The ship's deck cargo includes mainly timber, but also a section of a steel tower in the foreground

SOURCE Reproduced by kind permission of Scotline

Hull. All these routes operate on a shared user basis. They are also continually in the market for cargos to allow them to reposition ships to more favourable ports, given the one-way nature of the timber trade.

HMM, part of the Hyundai Group, offer liner services aimed at the oversize cargo market, from the Far East to the Persian Gulf, Mediterranean, Durban and South America (HMM, 2018).

These services can be accessed by contacting the shipping line directly, or through a broker or freight forwarder.

If there is no suitable liner service, or if you have a sufficiently large quantity of cargo to make it worthwhile, then the options of inducement or chartering (see below) should be considered.

Inducement

If you have freight that you wish to ship to a specific port, but that port is not served by a regular service, it may be possible to arrange for a ship to make an additional call at that port. This is known as inducement, which essentially means that you pay the shipping line enough additional money to make it worth their while to do so. Inducement is less common than it was, but is still possible.

It is recommended that you seek the help of an experienced forwarder or broker before taking this option.

Ship chartering

Most users of seafreight today will containerize their goods, or maybe ship as deck cargo or on a ro-ro service. However, if you are importing or exporting in large quantities, an option worth considering is to charter a dedicated ship. Examples might include 1,000 tonnes of forest products, agricultural products, or sand; 50 containers of components for car production; or 35 artic trailers of equipment for a major concert by a famous rock band.

There are two main types of charter – a voyage charter, whereby the ship carries a cargo from A to B; and a time charter, whereby you have control of the ship for a certain length of time, which might be seven days or seven years, and can send the ship wherever you want during that time within pre-agreed geographical limits.

It is possible to charter a ship directly from its owners, but I would not suggest this unless you are already experienced in this field. There are possible pitfalls and I would strongly recommend that you employ the services of a reputable shipbroker who can provide help and guidance.

Indeed, finding the right shipbroker can be the most difficult step in the process. There is a comprehensive directory called *The Shipbrokers' Register*, which lists over 11,000 companies worldwide. As of 2018 copies are €80 web only, €115 paper, or €125 for both (Shipbrokers' Register, 2018). In selecting a shipbroker look for experience in the type of ship you wish to charter, and possibly for qualifications from the Institute of Chartered Shipbrokers (MICS or FICS) amongst key staff. The Institute or the Baltic Exchange will also be able to name brokers from amongst their membership, and you can of course search the web – although you should beware of small operators claiming to be all things to all people. It is always prudent to seek personal recommendations from companies who have used a particular broker's services before.

The ship you charter will need to be large enough to accommodate the cargo, but small enough to meet restrictions for length, width, draught (depth) and air draught (height) in the ports being used. The smallest vessels available for charter are of approximately 1,500 tons deadweight, ie they can carry that weight of dense cargo, such as steel. At the other end of the market, general cargo vessels of 50,000 to 60,000 tons, or even bulk carriers of over 100,000 tons, are available, but if you are considering the charter of such large vessels, I would recommend that you recruit an experienced charter department manager.

For certain commodities, ships require specific certification. An example is grain, which could shift within the hold in severe weather, in extreme cases causing the ship to capsize. There are therefore regulations stipulating that there must be subdivision of the holds, and/or that a certain proportion of the grain must be bagged, to ensure that it can be carried safely. You should clearly specify the cargo you are seeking to transport, to avoid problems in this area.

Also, there is increasing regulation on the environmental impact of shipping. Maximum limits for sulphur emissions will be introduced worldwide on 1 January 2020 (IMO, 2018a) and even stricter limits are already in place in certain ports, including those in the Baltic and North Sea (IMO, 2018b). There are also regulations regarding ballast water. You should therefore ensure that the ship, and the fuel oil used, meet regulations for the ports in which you will be operating.

The terms of the charter are subject to negotiation. There are standard forms for what is known as a charter party, such as GENCON, which is often used for voyage charters; and BALTIME, often used for time charters. Samples of these can be found on the website of the Baltic and International Maritime Council. Changes can be made by mutual agreement of the parties but be careful to ensure clarity – contradictions between clauses can become a legal minefield. It is standard practice for all charter hire charges to be paid in advance.

In the shipping industry, the old principle of 'My word is my bond' still applies. Offers made and accepted are legally binding. If you ask 'Will you charter this ship to me for $10,000 per day?' and it is accepted, it is not possible to go back later and say 'I have had a better quote, will you do it for $9,500?'

There are numerous costs and charges relating to the operation of ships. Those directly related to the ship such as crew wages and meals, ship's stores, and maintenance, will almost always be paid by the owners. Other costs include:

- port dues;
- conservancy (which pays for keeping waterways navigable, eg by dredging);
- light dues, which pays for buoys, etc;
- pilotage;
- tugs (rarely used today as most ships are fitted with bow thrusters which enable them to manoeuvre themselves);
- bunkering (fuel);
- stevedoring (loading and unloading the ship).

For a voyage charter, it is normal to agree a cost for the voyage – this may be a fixed rate for the ship, or a rate per tonne subject to a minimum tonnage. The ship owner will then pay all the above costs, unless it is agreed otherwise: stevedoring is the commonest exclusion.

For a time charter, it is normal for the charterer to pay these costs. You will need to appoint a ships' agent in each port, who will organize payment of the first five elements on your behalf. They will almost certainly issue a proforma invoice, which you will need to pay by bank transfer before the vessel arrives at the port. The relevant port authority will provide a list of companies offering ships' agency on request.

> Beware of the use of the word 'free'. In shipping this means free to the ship owner, so 'free of port charges' on a voyage charter means that you have to pay them.

Regarding bunker fuel, it is normal for a charter party to include a clause which states that there is to be a specified minimum tonnage aboard at the beginning and the end of the charter, and that any difference in tonnage will be charged at the market rate as appropriate. If you are time chartering you will probably need to arrange bunkering, supplying the appropriate grade of fuel. This might be gasoil or heavy fuel oil, but liquified natural gas (LNG) is now also entering the market: Figure 14.8 shows *Engie Zeebrugge*, the world's first purpose designed LNG bunker barge (Engie, 2017), delivering the fuel to *Auto Eco*, the world's first pure truck and car

Figure 14.8 Bunker barge *Engie Zeebrugge* making an LNG fuel delivery to the car and truck carrier *Auto Eco*

SOURCE Reproduced by kind permission of Port of Zeebrugge

Figure 14.9 A ship on voyage charter. Below deck (not visible) is bagged cement, and loaded on deck are cement silos

SOURCE Reproduced by kind permission of Scotline

carrier able to run on LNG (Fairplay, 2016). You will again need to pay in advance, and delivery will most commonly be by barge, but sometimes by road tanker.

Stevedoring is an element that needs to be agreed with your customer in advance. Terms of sale will be discussed later, but if you agree to unload the ship in the port of arrival, either the ship owner will need to arrange this, which will be considered in their price, or more commonly you will need to arrange this. Again, the port authority will provide details of suitable companies on request, but you should ascertain the costs in advance.

If you are running an SME, ship charter will not be a viable option. However, if you are importing or exporting in large quantities, it is well worth considering.

Airfreight

Seafreight is the most popular mode by far for intercontinental shipments, and for some commodities is the only alternative: no one would seriously contemplate the airfreight of thousands of tonnes of iron ore. Airfreight is best suited to smaller consignments, and/or those for which speed is of the essence. As a result, although only about 3 per cent of the global trade by volume moves by air, it represents 35 per cent by value (Sales, 2017).

Examples of goods sent by air include:

- small consignments, which can often be sent more cheaply by air than by sea;
- perishable goods;
- fish from Scandinavia to Japan;
- roses from Ecuador to Europe;
- pharmaceuticals such as blood plasma, insulin and vaccines;
- fragile goods;
- goods urgently required for production;
- parts urgently required for repairs;
- animals, which would become stressed by a long sea voyage;
- very high-value goods, where the additional inventory cost of using seafreight can be a major issue;
- humanitarian aid, especially following natural disasters;
- those dictated by customer demand.

I will expand on some of these.

- When I was a child, food availability was dictated by the seasons. Strawberries were considered a short-lived treat in high summer, but today customer demand ensures that they are made available all year round. This practice has been criticized by environmentalists, who seek to reduce 'food miles', but this argument must be set against the genuine poverty that would be inflicted in what are already some of the poorest areas in the world if the trade in asparagus from Peru or green beans from Kenya were to be discontinued. Readers will no doubt differ in their opinions on the matter. Fruit and vegetables represent 8.8 per cent by weight of airfreight movements (Morrell and Klein, 2019).
- The just-in-time philosophy has reduced reserve inventory in production plants to a minimum. When I was in the automotive industry, the bane of my life was re-segmentation – a short-notice change in the mix of cars to be built on a production line (usually for good reasons, it must be said). This resulted in a change of the mix of parts required, often in turn requiring airfreight. Today there are an average of 10 charter flights per night in Europe for automotive parts, eg from Poland to the UK.
- I mentioned in earlier chapters that excellent roadfreight networks exist to serve the vehicle off road sector of the automotive industry. However, the cost of downtime due to breakdown for other items can be frightening – an airliner may be unable to fly, a Formula One car unable to race, or an offshore rig unable to produce oil or gas. In such cases the cost of airfreight will be money well spent.

- For very high-value goods, the cost of inventory needed to support a seafreight-based supply chain of several weeks will be substantially greater than that to support an airfreight-based supply chain of several weeks. This will affect the balance sheet, and in extreme cases the cost of capital to support that inventory will exceed the additional cost of airfreight.
- Finally, there is the question of customer demand. Consumers in China, for example, increasingly demand the latest styles from the great fashion houses of Europe (Sales, 2017). These must be flown in to meet that demand – by the time they arrive by sea they will no longer be the latest style. The same principle may apply to the latest video game, and in some cases the customer will be prepared to pay the additional cost of airfreight simply because they are not willing to wait. Finally, some goods have a very brief selling window. For example, in the UK Halloween is a short annual retailing season, and goods relating to major football (soccer) tournaments suddenly cease to have any value on a particular team's exit from the competition.

How airfreight works

The airfreight industry can be subdivided into the charter market, which I will discuss in a later section, and the scheduled sector.

Within the latter, there are of course dedicated freighter aircraft, but at an airport such as Heathrow these represent less than 1 per cent of aircraft movements. Most freight is carried in the belly holds of passenger aircraft. Modern wide-bodied planes have very large cargo compartments beneath the passenger seating, far more than is required for accompanied luggage, and cargo provides 19–20 per cent of total revenue for Emirates Airline, and about 33 per cent for Cathay Pacific (these figures include dedicated freighters). Many low-cost carriers, however, have a policy of not entering the air cargo market (Sales, 2017).

There are 8,707 airports across the world. Of these, some handle a higher proportion of freight than of passengers, such as Kansas City and Paris Vatry (Sales, 2017). A number have heavy peaks of demand, such as Chateauroux, the export gateway for much of each season's Beaujolais Nouveau wine.

In many cases, initial movement of airfreight will be by truck to a different airport that has a direct link to the destination, for example from Heathrow to Paris Charles De Gaulle or Amsterdam Schiphol.

On arrival at the airport of dispatch, goods are subject to a 100 per cent security check (unless they originate from a shipper authorized to carry out their own checks). They will probably be loaded into what is known as a unit load device. These are similar in principle to a shipping container and shaped so as to fit into the aircraft. Alternatively, larger cargo may be loaded directly. A variety of loading equipment is used, including conveyors and scissor lifts (Figure 14.10).

Figure 14.10 Scissor lift being used to load time-critical automotive parts aboard a Boeing 747-400F aircraft for shipment from London Stansted to Bari

SOURCE Reproduced by kind permission of Chapman Freeborn

The cargo may then be routed via one or more hubs – for example a shipment on American Airlines from London to Mexico will be transhipped at Dallas/Fort Worth (Sales, 2017). On arrival at destination, customs clearance must be completed before goods are released for delivery. Well-prepared paperwork will help to avoid delays at this stage.

> Airlines are well known to oversell tickets for passengers, due to an anticipated proportion not booking in at the airport. This sometimes results in passengers with a valid ticket being 'bumped off' and not permitted to fly because of insufficient space. The same principle applies to freight, and there is a risk of goods being bumped off in the same way. This is a reality you should be aware of.

Access to scheduled airfreight services

> To access these services, you need to use a freight forwarder as an intermediary, or you can use what are known as integrators. This in practice makes matters much easier for the inexperienced user.

Most airlines do not have large customer service departments in their cargo operation, nor an accounts department able to accept payment from a plethora of sources: they rely instead on the Cargo Accounts Settlement System used by forwarders worldwide. The use of freight forwarders to access freight services from scheduled airlines is therefore almost universal.

You will be able to find a freight forwarder through the BIFA website or overseas equivalents such as FENEX in the Netherlands. They will arrange trucking to collect the consignment from your premises and deliver to final destination, produce paperwork such as the air waybill, and arrange for customs clearance.

Integrators are so-called because they provide an integrated range of services in-house, including a fleet of dedicated freighter aircraft. The big three are DHL, UPS and FedEx. Each of these will allow you to book a consignment online, paying by credit or debit card if you are making a one-off booking. It is easier to open an account online if you are making regular shipments, even for an SME. They will also arrange trucking and customs clearance, and provide online tracking systems.

My main advice if you are shipping regularly is to obtain a telephone number for someone at your local depot. If there is a problem, a centralized call centre may not offer the level of personalized service that you would like, and I personally found it reassuring to be able to talk to a familiar person on the rare occasions I needed to do so.

Your responsibility is to provide full and accurate information. Some people are tempted to under-declare consignment weights and dimensions to obtain a cheaper price. This can however be dangerous, in that an aircraft can be overloaded or unbalanced, and can be used as an excuse by some customs authorities to impose heavy fines for providing false information, delaying the consignment until the fine is paid (in some cases failing to inform you and charging interest for late payment!).

Full and accurate descriptions of goods, and unambiguous delivery address, are also vital. Any supporting documentation required, such as certificates of origin (see Chapter 17), must also be in good order to avoid delays.

Dangerous goods

There have been serious accidents, and people have died, due to failure to accurately declare dangerous goods.

Incidents have included undeclared bottles of nitric acid, believed to have ignited other material and caused an aircraft to crash into the Indian Ocean; and flammable liquid leaking from a passenger's baggage, causing the plane to crash and the deaths of all on board. Near misses include fuel leakage from fuel tanks of mowers and chain saws, and an undeclared oxygen generator that burst into flames whilst awaiting loading (Sales, 2017).

Your main responsibility is to declare all dangerous goods openly and honestly, and comply with regulations for packaging and other requirements. Remember that perfumes, blood for transfusion (potentially containing pathogens), and toys fitted with lithium batteries can be dangerous – if in any doubt you should put the question to your forwarder or integrator.

The International Air Transport Association (IATA) issue a comprehensive manual, *Dangerous Goods Regulations* – the 60th edition came into force on 1 January 2019. It can be purchased online, but costs several hundred dollars: whether you should buy a copy will depend on your individual circumstances. They also arrange courses on the subject.

Animals

The transport of animals by air is now routine airline business. This includes domestic pets accompanying passengers, exotic pets such as reptiles and live fish. The transport of zoo animals such as hippopotamuses may require several months of planning to ensure that their crates are designed to prevent escape but not to stress the animal – a hippo loose at 36,000 feet might sound funny but would not appear that way to those on the plane.

Racehorses are usually accommodated on charter aircraft, and accompanied by a groom – who will need security clearance to be permitted to fly on a freighter.

There are quarantine restrictions in some countries to control the spread of diseases such as foot and mouth. For example, racehorses arriving in Australia will be detained for two weeks after arrival (Sales, 2017). This should be taken into account in planning – arriving only a week before a planned horse race would be an embarrassing error.

IATA also issue regulations for the transport of animals, available to purchase online.

Charter aircraft

Most readers will be familiar with this term and will have at some point taken a charter flight as a passenger as part of a package holiday, the holiday company having chartered the whole aircraft to take its customers to their destination. The same principle applies to cargo aircraft – you can charter a complete aircraft to take your goods (and only your goods) from A to B.

The market has changed in recent years. There are fewer aircraft available for charter – many older Boeing 707 and Belfast aircraft are no longer flying, and some do not meet European noise regulations – but more charters are being fixed. As a result, it is useful to plan as far ahead as possible to have the best chance of finding a suitable aircraft available. There is also a higher degree of regulation, for example to prevent bribery.

The main reason for using charters is assurance. Using scheduled services, cargo can be bumped off a flight, and connections can be missed. Charters are dedicated to your needs, and can fly directly between airports of your choice, so these risks are avoided. Also, some very large items can only be accommodated on charter aircraft. Figure 14.11 illustrates all these criteria: there is no direct scheduled flight from Glasgow to Luanda, a delay to the equipment would no doubt cause major problems, and its size would preclude use of a scheduled service. Major users of charter include the automotive and defence industries, and emergency humanitarian aid organizations.

At the large end of the market, the Boeing 747-400 has a payload of 110–120 tonnes, and a very long range. However, they are not easy to charter at short notice.

Figure 14.11 A Boeing 747-400 aircraft being loaded with outsize oil and gas equipment for a flight from Glasgow to Luanda

SOURCE Reproduced by kind permission of Chapman Freeborn

Other aircraft available on the charter market include:

- Antonov 12, approximate payload 20 tonnes;
- Antonov 74, 10 tonnes;
- ATR 72, 8 tonnes;
- BAe 146, 8 tonnes;
- Antonov 26, 5.5 tonnes (Figure 14.12);
- ATR 42, 5 tonnes;
- SAAB 340, 2–3 tonnes;
- Metroliner II or III, 2–3 tonnes.

The smaller aircraft tend to have a range of two to three hours flying time, but refuelling stops can be made if necessary.

It is also possible to charter helicopters for very urgent movements into areas that cannot be reached by fixed wing aircraft: I have done so on very rare occasions for deliveries directly into automotive plants. The Russian-built Mi 8, with a payload of 4 tonnes, is the largest available, but usage on charter is in practice minimal. Indeed, most helicopter charters are for passenger use, such as businessmen attending meetings, rather than for freight.

Figure 14.12 An Antonov 26 aircraft, one of the smaller types available for charter, with a payload of 5.5 tonnes

SOURCE Reproduced by kind permission of Chapman Freeborn

Access to air charter services

It is possible to charter aircraft directly from airlines, but I would recommend using a charter broker such as Chapman Freeborn. They will have built up a pool of knowledge as to which airline is likely to have a suitable aircraft available at that time, and which would be the best airport to use. To obtain a landing slot at Heathrow for an ad hoc charter would be prohibitively expensive, and in practice impossible; a small airport might not have the necessary handling facilities. Charter brokers will also be familiar with regulations such as pilots' hours, which are far more complex than those for HGV drivers. This knowledge will be invaluable, especially to the inexperienced charter user.

For a new user, payment will be required in advance. This is usually by bank transfer, but credit or debit cards may also be accepted.

On-board couriers

For small consignments requiring an exceptionally urgent and dependable delivery, the use of an on-board courier is a possible option. This is sometimes known as a freight-in-baggage service. Examples of such consignments might include a component required for a repair, or samples needed for a vital sales presentation.

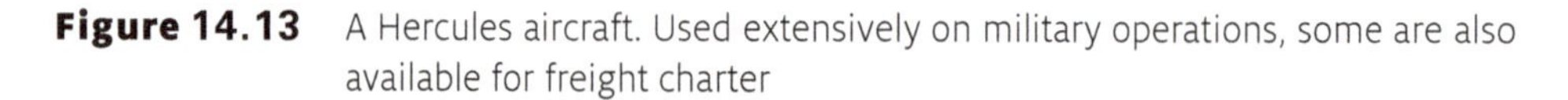

Figure 14.13 A Hercules aircraft. Used extensively on military operations, some are also available for freight charter

SOURCE Reproduced by kind permission of Chapman Freeborn

The service involves the courier taking a scheduled flight as a passenger and carrying the item in their baggage. An agent will need to be appointed at the arrival airport to arrange customs clearance. The courier may be met at the airport by the consignee, or take a taxi to the destination.

This service is naturally very expensive, but when time is more important than cost it can be extremely useful.

Drones

Recent publicity has suggested the use of drones for urban deliveries. For isolated consignments, this may be feasible. However, it is unlikely that drones will replace vans for routine parcel deliveries for many years – the technical difficulties of avoiding collisions between drones or with obstructions, and crashing and injuring people below, will not quickly be resolved. Drones do seem successful in one type of delivery, that of drugs into prisons, but I do not predict widespread use in the foreseeable future.

Sea–air services

There are circumstances in which seafreight is too slow, but you do not need quite the speed of airfreight. There is a compromise available called a sea–air service. If, for example, you are shipping from Japan to the UK, you could ship the goods by sea to Dubai or Seattle, and then use airfreight for the remainder of the journey. The service is sometimes said to be half the price of airfreight and half the time of seafreight, but a more realistic assessment might be two-thirds of the cost of airfreight and two-thirds of the time of seafreight. The China-to-Europe rail service mentioned in Chapter 13 would be another alternative.

Sea–air services might, for example, be used as part of the launch plan for the latest mobile phone. The first consignment might be sent by air, the second by sea/air, and the third by sea, all leaving in quick succession but perhaps arriving two weeks apart.

Most supply chains are set up to depend either on airfreight or seafreight, but occasionally a compromise is needed, and it is reassuring to know that one is available.

Conclusion

This has been a wide-ranging chapter. I have described seafreight movements, from the routine use of ro-ro and container services by an SME to the charter of a ship for very large consignments; and airfreight, a faster method of long-distance transport, but more expensive, especially if chartering an aircraft or using an on-board courier. Together with previous chapters, I have now covered all major transport modes – road, rail, inland waterway, sea and air.

I will now move on to other matters affecting logistics, beginning in the next chapter with terms and conditions, liability and insurance.

Trading terms and insurance 15

The commonest lie on the internet is 'I have read and agree to the terms and conditions.'

The above statement was made by security expert Mikko Hyppönen (*Guardian*, 2015), and I am sure he is right. When agreeing to a transaction, such as booking a freight movement, it is very tempting to tick the box agreeing to terms and conditions without reading them and ensuring that they are fully understood and contain no sinister clauses. There can be few of us who have not been guilty of this at some point, but taking the time to study, and if necessary challenge, standard clauses can save a lot of heartache should something go seriously wrong.

In this chapter I will look at four aspects of this topic.

- incoterms, sometimes called freight terms;
- payment terms for international trade, in particular letters of credit;
- the more common standard terms used, such as CMR and RHA;
- insurance – how costs can be mitigated whilst avoiding undue risk of a catastrophic financial loss.

Incoterms® – who pays for which parts of the freight cost?

Incoterms® is a registered trademark, and an acronym of INternational COmmercial TERMS. These are a set of rules issued by the International Chamber of Commerce (ICC) to offer a choice of selling terms, in relation to freight costs and liability, and to standardize rules within those options. It is important to mutually agree which term will apply, and for both parties to confirm this to the other: contradictory terms create innumerable problems. This is especially important if the two parties wish to deviate from the standard terms.

As at 2019 there are 11 options available, but I will home in on those encountered most commonly:

- Ex-works (EXW). This means that the buyer is responsible for sending a vehicle at their own expense to the seller's premises to collect the goods. The seller is responsible for packing and labelling the goods, and for issuing a commercial invoice (Croner-I, 2018), but nothing else. In practice, this is one case where I would recommend a pre-agreed deviation from the standard EXW terms, that the seller should load the goods onto the vehicle at their own expense and risk. It should be noted that the buyer is responsible for arranging customs clearance and other export formalities, and that domestic sales taxes such as VAT will be payable (Croner-I, 2018).
- Free carrier (FCA). This means that the seller is responsible for delivering to a carrier's premises – the exact location should be agreed in advance and the terms expressed in full, for example FCA XYZ Airfreight Services Heathrow. The seller will also be responsible for packing, labelling, and export formalities. Terms sometimes state FCA (seller's premises) which may sound illogical but overcomes the problem that under the strict interpretation of EXW the seller is not obliged to load the goods onto a vehicle.
- Free alongside ship (FAS). This should only be used for transport by sea or inland waterway. The seller is responsible for freight charges to the relevant port, but the buyer is responsible for the cost of loading the goods onto the ship and all costs thereafter.
- Free on board (FOB). As FAS, but seller also pays the loading costs.
- Cost and freight (CFR). This is also for waterborne freight only. The seller pays for the seafreight in addition to FOB costs, but the risk during the seafreight leg rests with the buyer. Also, there is no standard rule regarding handling charges at the port of arrival – this should be agreed in advance. Import customs clearance, duties and taxes, and transport to the buyer's premises are costs for the buyer.
- Cost, insurance and freight (CIF). As CFR, but the seller must insure the goods. However, risk in the goods during the sea passage remains with the buyer, and no particular level of insurance is specified, so if insufficient care is taken it is possible to find the goods greatly under-insured and for the buyer to suffer a substantial loss as a result.
- Delivered duty paid (DDP). The seller pays all costs to deliver the goods to the buyer's premises. The only cost for the buyer is that of unloading the vehicle. This is at first sight the easiest option for the buyer, but in practice a seller may not have the wherewithal to arrange this, and the buyer may wish to be in direct control of the shipment once it reaches the destination country.

As mentioned before, there are options within some of the terms, and they are silent on some matters, such as the responsibility for stowing goods correctly within a container, and at which point title in the goods passes to the buyer. Imprecision is another common issue, with terms such as FCA New York – New York is a big place!

Also, a revised set of rules will come into force on 1 January 2020. The ICC publish guides to the current rules and I am sure they will publish similar guides to Incoterms 2020.

Bearing all this in mind, I would recommend a detailed written agreement between both parties as to what exactly each party will be paying for – re-iteration of basic terms; clarification on unspecified matters; and levels of insurance where applicable. In this way you will be able to ensure that there are no nasty surprises for either party.

Payment and letters of credit

Strictly speaking, these are matters for the sales and accounting functions of a business. However, it is useful for a logistics manager to have a knowledge of the requirements for international trade, and I will therefore provide an overview.

The main options are as follows.

Payment in advance

This may be by bank transfer or by debit or credit card. For small transactions, for example if you are selling to overseas consumers, this will be the preferred method, and is obviously the most advantageous to the seller in terms of both cash flow and certainty of payment. However, if you are selling to commercial buyers they may be unwilling to accept this option, and this will place you at a competitive disadvantage.

Open account

This is the method used for most domestic commercial transactions – an invoice is issued on supply of the goods, and payment must be made within a specified period (eg 30 days). There is a risk of non-payment, whether due to the bankruptcy of the buyer, a real or alleged problem with the delivery, or simple dishonesty. This will be the most attractive method to the buyer, and the risk to the seller of non-payment can be assessed by considering the financial stability, reputability and past payment record of the customer to establish their creditworthiness.

Letter of credit

> The basic structure is very very simple. An undertaking to do something (broadly, to render banking services) at the request of the account party in favour of the beneficiary party provided certain conditions are met (conditions precedent or CPs). Nothing else.
>
> (Kurkela, 2008)

The above is a succinct summary of letters of credit. Usually, it will contain a commitment from the buyer's bank to make a payment to the seller's bank when goods are delivered, and this has been satisfactorily evidenced by the provision of pre-agreed documentation. They are most often used where the buyer has poor creditworthiness, or it is difficult to establish whether they have or not. If the buyer should enter bankruptcy, or default on payment, the bank will be obliged to make the payment instead.

Their main advantage to the seller is that the bank is much less likely to default on payment than the buyer, so in simple terms the chances of not receiving payment are much reduced. The advantage to the buyer is that their apparent creditworthiness is much increased by the commitment from the bank, and this will often in practice be the only way in which they can obtain credit.

Great care must be exercised in the preparation of letters of credit, as small errors can be used as loopholes to avoid payment. If you are inexperienced in their use, I recommend that you seek professional help in drafting them.

They can also be expensive, with charges from banks amounting to thousands of dollars in some cases, and this should be taken into account when calculating the selling prices for your goods.

Common pitfalls for sellers include:

- Making payment conditional on an action by the buyer. For example, if the letter of credit says that payment will be made after the buyer issues a proof of delivery, an unscrupulous buyer can accept the goods, and avoid payment by deliberately not issuing a POD.
- Describing the goods in very fine detail, giving more opportunity to find a discrepancy.
- Not asking to see a copy before the letter is sent to the buyer's bank, and/or not raising any necessary amendments quickly.
- Overlooking the time needed to manufacture and deliver the goods.
- Ignoring the need for everyone's approval with everything in the document – a situation where the production manager states 'I can't do that' after documents have been issued will be a big problem. Everything should be checked and rechecked.
- Being unable to provide all the documents you agree to provide, without excessive time, trouble or expense.
- Not involving your freight forwarder in the preparation process – they may be a vital link in the chain.
- Involving a third party in the agreement, for example by committing the buyer to payment only when they have been paid for the goods.

Whilst care in preparing documents is always important, it is especially so with letters of credit. 'Near enough' is *not* good enough (Chamber International, 2018c).

Further reading

The International Chamber of Commerce issues the *ICC Uniform Customs and Practice for Documentary Credits UCP 600*, more commonly known simply as UCP600, which details standard rules for letters of credit. For more advanced reading, *Letters of Credit and Bank Guarantees Under International Trade Law* by MS Kurkela (2008), published by Oxford University Press, provides a comprehensive discussion from the legal viewpoint.

Consignment stock

Another alternative is to appoint an agent or distributor in the overseas country. The distributor holds the goods in their warehouse but will only make payment to you when they have received payment for the goods. This carries a higher risk than other options of never receiving any money, and I would only advise entering into such an agreement if you are very sure of the reliability of the distributor.

The CMR convention

The *Convention Relative au Contrat de Transport International de Marchandises par Route*, more commonly known as CMR, was signed in 1956, and incorporated into UK law in 1965 (Lowe and Pidgeon, 2018). The title translates as the Convention Relating to the Contract for the International Carriage of Freight by Road. It governs all movements (with limited exceptions) between over 45 countries, including most of Europe, Morocco, Syria, Jordan and some constituent countries of the former USSR. It applies automatically, even if the carrier is unaware of it or attempts to opt out by asking the consignor to sign documents to that effect. It does not, however, cover the carriage of your own goods on your own vehicles.

You can buy pads of forms online, or buy software to print them. The form should be completed in full for each consignment, and all four copies of the CMR set should be signed by sender and carrier (usually in practice the driver). Of the copies, one is retained by the sender; one accompanies the goods and is ultimately handed to the consignee; one is retained by the carrier and remains on the vehicle until the goods are delivered; and one is sent to the carrier's office for retention.

The carrier is responsible for checking the goods, to ensure that the packages and marking match the CMR forms, and that there is no apparent damage. There is a box on the form in which the carrier can enter 'reservations', such as boxes appearing damaged or wet. Comments such as 'Not examined for damage' should not be accepted: the goods should be examined.

The carrier is responsible for loss or damage to goods whilst in their care, unless caused by wrongful act or neglect of the claimant, for example damage due to insufficient packaging. The burden of proof rests with the carrier.

Liability is limited to 8.33 special drawing rights (SDRs) per kilogramme of goods. The SDR is based on a basket of currencies, and its value relative to individual currencies varies day by day. Its value can be found on the internet, and is printed in the UK in the *Financial Times*. To give some idea of its value, as of 17 December 2018 one SDR was valued at £1.09, US$1.38 or €1.22 (IMF, 2018).

Other terms and conditions

If you are a large user of logistics services, it will be possible to insist that carriers adhere to terms and conditions that you have specified – qualified legal assistance should of course be sought in drafting these.

However, for most users, the reality will be that the terms and conditions specified by the carrier will have to be accepted. These may be unique to that carrier, but in most cases they will use standard conditions issued by a trade body. These will usually include some terms to cover obscure points of law. For example, in the UK it is usual to specify that a haulier is 'not a common carrier'. A common carrier has many onerous duties, including being unable to turn away unwanted customers (Kahn-Freund, 1965). I know of no company still trading as such, nor can I conceive of a reason for anyone wanting to do so. Such terms are of little practical importance to the service user.

Most domestic road hauliers in the UK use the Road Haulage Association (RHA) Conditions of Carriage 2009. These are not in any way unreasonable, but it is wise to be aware of their implications. For example, liability for loss or damage is limited to £1,300 per tonne (Armstrong Logistics, 2018). This will cover some bulk products such as coal, but even materials such as rough sawn timber are likely to exceed this value. Any manufactured goods will almost certainly do so. Furthermore, events beyond the carrier's control, such as poor packaging, Acts of God, or riots are excluded. Unless you are prepared to accept these risks, most customers will therefore require additional insurance, either by agreeing a higher level of liability with the haulier and paying additional charges to cover this (beware that this does not still exclude risks such as Acts of God), or by arranging insurance directly with a broker or insurer. You should also make yourself familiar with time limits for raising claims.

For storage, both RHA Conditions of Storage 2009 (Armstrong Logistics, 2018) and United Kingdom Warehousing Association Contract Conditions for Logistics 2014 (PD Ports, 2018) specify a default liability limit of £100 per tonne. This would be unlikely to be sufficient even for a commodity such as scrap metal, so in practice you will need to arrange additional cover with the warehouse keeper or with an insurer.

Many freight forwarders will operate under the standard terms of their national association, such as BIFA in the UK, or FENEX in the Netherlands. Whichever is the case, my main recommendation is to make yourself aware of those terms and arrange whatever additional cover is necessary for your particular business.

Insurance

My personal view of insurance differs from that of some others. In my view, its purpose should be protection against major losses, through whatever cause, that will have a significant impact on the business. Its purpose should not be to enable you to make frequent minor claims.

The reality is that insurance is a business, and all businesses must make a profit in order to survive. If an insurer is aware from past experience that they will be paying out, say, £10,000 in minor claims each year to your organization, they will factor this into your premium in future years, plus a profit margin, and charge a further premium to cover major risks on top of this. A better approach in the long run is to cover small losses yourself and strive to minimize these by good management. Accepting a higher voluntary excess on your insurance policy will reduce premiums, and ultimately reduce overall expenditure.

One major difference between private and commercial insurance is a responsibility to volunteer information. For the former, you must answer all questions honestly, but are not usually obliged to go further. However, for the latter, you must disclose any relevant facts you know, or should have known if you had made reasonable enquiries. If, for example, you do not declare that your goods are hazardous or attractive to theft; or that there is a history of flooding in the area, and goods in your warehouse are damaged by a flood, your cover may be invalidated, or pay-outs reduced.

The list of possible risks against which you can insure is almost endless. Few businesses, especially small or new businesses, can afford cover for them all. Apart from legal requirements, you will need to assess the risk of an event occurring, the cost if that event should occur, and the premium to guard against it. Your own personal attitude to risk will be another factor, and you can then decide upon the balance of risks that you do and do not wish to protect against. An insurance broker will be able to help you decide which policies are right for your particular business. For a very large company a major organization may be the best option, but for a small to medium sized business a smaller broker such as Marlow Gardner and Cooke may be more appropriate. The British Insurance Brokers Association or Chartered Insurance Institute may be helpful in identifying a suitable broker.

Types of insurance available

All businesses would be well advised to take out certain insurance policies, for example buildings and contents insurance on their offices, public liability, and employers' liability insurance – the latter is of course a legal requirement.

Some other types of insurance policies available on the market are particularly relevant to logistics operations. They include the following:

- Property insurance on the warehouse, to include:
  - equipment such as forklifts;
  - external units such as fuel storage tanks.
- Cyber insurance, eg to cover an attack on your warehouse management system and loss of data.
- Business interruption, eg against losses due to destruction of the warehouse and inability to trade.
- Motor insurance – third party, third party fire and theft, or comprehensive:
  - gap insurance will cover the difference between the actual value of a vehicle, which is what would normally be paid in the event of a write-off, and the sum payable under a lease or hire purchase arrangement, which will almost always be greater.
- Goods in transit insurance, to include goods in storage. It is unwise to rely on cover provided as standard by a carrier (see above):
  - goods should be covered to their full value;
  - your own annual policy is likely to be more cost-effective than per consignment cover offered by a carrier;
  - crime insurance, for example against a claim because one of your employees aids a theft or fraud, is available.

- Credit insurance, to cover bad debts when trading, domestically or internationally.
- Legal expenses insurance:
  - property, eg against boundary disputes;
  - motor, to defend against prosecutions or pursue uninsured losses;
  - management liability, in case managers or directors are sued personally.

There are a few specific points I will mention.

- If you carry a forklift on your vehicle, and it is used to unload the vehicle in a public area (such as the service road behind a row of shops) it will require insurance under the Road Traffic Acts. If it is uninsured and injures someone, the repercussions will be very serious.
- Ancillary equipment on vehicles, such as cranes or tipper bodies, must be declared to insurers, as must the nature of any trailers.
- Similarly, you must declare operations in high-risk areas such as ports, rail-heads and nuclear installations. Most policies are not valid in airside areas of airports.
- If you are insuring goods in transit by seafreight, ensure that risks under General Average are covered. This is a little-known facet of maritime law, under which costs of salvage or rendering assistance to a ship in distress (eg tugs or deliberate jettisoning of cargo) are shared amongst both the ship owner and the owners of cargo. If you have goods in a container on board you will have to pay part of the costs.

In Chapter 10, I discussed accident prevention and ways to mitigate costs. It will help to reduce your premiums if you can show that you have procedures in place to do so: you may have to wait one or two years to show that the actions have been successful, but they will benefit you in the long term. Measures likely to be attractive to insurers include: four-way cameras on vehicles, especially in cities; enhanced security, both on vehicles and at parking places; not parking vehicles closely together – this increases the risk of damage to numerous vehicles in the event of an arson attack; good defect reporting and rectification records; driver vetting and training; and willingness to accept a higher excess – this may be compulsory in some circumstances, such as inexperienced drivers.

Conclusion

There is no such thing as a risk-free business. Often, the key to a successful business lies in managing those risks – assessing likelihood and costs, and accepting, mitigating, managing, avoiding or insuring against them according to your personal best judgement. A key element in this process is agreeing the most appropriate terms of business with your customers and your suppliers.

This chapter explored four core aspects – freight terms, payment terms with customers, terms and conditions used by logistics providers, and insurance. There are no right or wrong answers, but the information provided will help you to judge the best answers for your particular business.

Customs formalities and imports

16

Customs formalities are an immensely complex subject and vary from country to country. In practice, most companies use an agent to handle declarations on their behalf, even major multinationals. Unless you already have a thorough knowledge and understanding of customs matters, I would recommend this option.

Attempting to deceive customs authorities is a bad idea. Not only can they check back for a period of up to seven years and impose large retrospective penalties, but they are likely to repeatedly select your goods and paperwork for extensive checks, with long delays and possible costs as a result. Honesty is by far the best policy.

Having said this, a basic understanding of the processes is useful, and there are some ways in which you can quite legitimately delay, or even avoid, the payment of duties. In a book entitled *A Practical Guide to Logistics*, advice on reducing the cost of your imports and/or improving cash flow by doing so is perhaps the most important, so I will be devoting the latter half of the chapter to some of those methods.

Before doing so, I will however first explain some of the underlying principles.

Systems: CHIEF, CDS and CPC codes

Customs Handling of Import and Export Freight (CHIEF) is the long-standing protocol for passing information to and from UK customs. The original basis was form C88, the Single Administrative Document, which was standardized throughout the EU and other European countries. The form has 54 boxes, although not all of these require completion for any one transaction. In theory it is still possible to use a paper form for some actions, such as applying for export clearance, but in practice not all ports still have an Export Processing Unit and there is therefore no customs officer to whom you can physically hand the form. Therefore, I do not recommend this.

Your agent will have access to Her Majesty's Revenue and Customs (HMRC) online, and will create a substitute electronic version of the C88, with the relevant information that would have appeared in the relevant boxes.

However, CHIEF is in the process of being phased out in favour of the Customs Data System (CDS). As of 2018 CDS is in use for customs warehousing (see below), and will become available for imports and exports during 2019. The two systems will operate alongside each other during the launch period: a final date for the complete discontinuation of CHIEF will be announced in due course, and will probably be during 2020 or 2021. There will be differences in the way in which information is presented, which will be based around eight data sets rather than the 54 boxes. Some additional information will also be required; for example, buyer and seller must clearly be specified.

Customs Procedure Codes (CPCs) are seven-figure numbers to indicate what type of customs transaction is taking place. For example, goods in free circulation for export outside the EU will usually be processed under CPC 10 00 001 (there are exceptions such as alcohol and tobacco). CPCs, of which there is a very long list, should not be confused with Harmonised System codes (see below). Your agent will be able to provide the appropriate code for whichever particular procedure you wish to follow.

Customs Notices

HMRC issues a series of what are known as Customs Notices, of which over 60 relate to imports and exports. Several of these will be referred to in this and the next chapter.

They are most easily viewed online and are in a user-friendly question and answer format, avoiding technical jargon where possible. Many are very specialist. For example, Excise Notice 162 relates to cider production. This will be very important to those involved in this business, but not to anyone else. I would recommend familiarizing yourself with any Customs Notice related to an activity with which you or your organization become involved.

Economic Operator Registration Identifier

The first thing that all businesses involved in imports and exports will need is an Economic Operator Registration Identifier (EORI). This replaces what used to be known as a Trader's Unique Reference Number (TURN).

You should apply online via the Gov.uk website, and you will normally receive your EORI within three working days by email. Your agent will need to be told what it is before they can clear goods on your behalf.

If you have a VAT number (eg 987654321) this will be used as the basis for your EORI, and if you are UK-based it will be GB 987654321 000. If you are not VAT registered, you will be allocated a number that resembles a VAT number.

HS codes: What is the rate of duty on this item?

The short answer is that it depends what the item is. Before you can answer the question, you need to work out the commodity code. There is an internationally agreed set of codes called the Harmonized System (HS). It is important to note that although all nations use the same codes, they do not all charge the same rates of duty.

Most governments provide an online database that allows you to search for items; for example, the UK Government (2018d). A full code has 10 digits, of which the first four are known as the heading code. As an illustration, let us assume that you are importing a burglar alarm to install in your new warehouse. A search for 'burglar alarm' lists the following:

85 Electrical machinery and equipment

 85 31 Electric sound or visual signalling apparatus

  85 31 10 00 Burglar and fire alarms

  85 31 10 30 00 For use in buildings

  85 31 10 95 10 For use in civil aircraft

  85 31 10 95 90 other [eg for use in cars]

As the alarm is to be installed in a warehouse, HS code 85 31 10 30 00 will apply. The website states that the duty rate for importing this item into the UK is 2.2 per cent.

A search of the corresponding website for New Zealand (New Zealand Government, 2018) shows the same eight-digit HS code, 85 31 10 00, but with a standard duty rate of 5 per cent for all such alarms.

In some cases, it will not be easy to search the website – for example you may have developed a new type of pre-prepared food product with a variety of ingredients, or you may be in doubt: is an item a toy or educational? Indeed, many people who are not familiar with HS codes find the concept difficult to understand at any time. If you wish, you can contact a helpline by email or by telephone – Customs Notice 600 will have the current details. If you require certainty, you should seek a written Binding Tariff Information decision, which will be binding on both you and customs authorities for three years. It is wise to determine the code and relevant duty rate in advance, both to avoid potential delays due to incorrect information, and to consider when calculating the selling price for your product.

One point to note is that VAT is payable on the value of the goods after payment of duty. Consider, for example, a pair of adults' shoes (children's shoes are VAT exempt), valued at £50, which attracts import duty of 8 per cent. The overall cost will be:

Base price		£50.00
Duty	8 per cent	£ 4.00
Sub-total		£54.00
VAT	20 per cent	£10.80
Total		£64.80

You will also need to quote the HS code on your commercial invoice (see Chapter 17) if you are exporting to some countries.

In addition to import duty and VAT, some goods – alcohol, tobacco and petroleum oils – also attract excise duty. This must be paid before the goods can clear customs.

At the other extreme, some items, such as books, are duty and VAT free.

Buying the customs tariff

It is possible to buy a copy of the UK customs tariff from the TSO Shop (the successor to Her Majesty's Stationery Office). It comprises three volumes:

1. Information including duty relief schemes, contact addresses and explanation of subjects such as Excise Duty and tariff quotas.
2. Duty rates and descriptions on 16,600 commodity codes.
3. Guidance on customs procedures.

It comes in four formats: online, CD-ROM, E-Tariff (a data feed in ASCII format), or printed. The latter is supplied in three large loose-leaf binders, and such is the size of the publication that some people find it physically difficult to carry all three binders at once. As at 2018 a print copy costs £389, including an annual subscription to monthly updates (HMRC, 2018b).

Import clearance: How it works

The process will start when you make a pre-advice to your agent. Normally this will be required at least two clear working days before the arrival of the goods, and you should send the agent the bill of lading, air waybill, or road vehicle documentation; and the commercial invoice (or a copy, if originals are accompanying the goods). The agent will then raise a pre-advice on CDS or CHIEF, and submit this to customs.

Assuming the goods are on a truck, the driver should report to the agent's office on arrival at the port. The agent will notify customs electronically. After a specified period (typically one hour) a response will be received from customs. This may either be confirmation that the goods have been cleared, or a notification that they wish to inspect the goods. In the latter case the driver will need to proceed to an inspection area, and the inspection will be carried out – this may be superficial or extremely thorough and time-consuming, at the absolute discretion of the Customs Officer.

On completion of the process the vehicle can leave the port, and the agent will send you documents confirming clearance. You can agree with your agent whether this is done individually or in bulk, perhaps weekly or monthly.

Goods that have been customs cleared with all duties and taxes paid are said to be in free circulation. They can then be taken anywhere within the relevant customs territory (eg the EU Customs Union) without further formality.

Export clearance

This works in a similar way. You should make a pre-shipment advice through your agent in the port of exit, which will need to include:

- exporter details, including VAT number;
- agent details, including VAT number;
- consignee;
- description of goods, plus quantities, weights and dimensions;
- origin of goods;
- HS code;
- CPC code;
- value of the consignment;
- unique consignment reference (UCR).

Your agent will normally provide details such as the CPC code – few shippers will know all the options available.

The UCR is used to identify the particular shipment. The recommended format consists of:

- year (two digits), eg 18 for 2018;
- country of export, eg GB;
- your EORI (see above);
- a unique number for that specific consignment.

Therefore, the full UCR might read 18GB987654321000-J12345.

An arrival message must be presented to customs when the goods arrive at the port. Customs officials will sometimes choose to examine the goods, but this is less common than for imports. Clearance will in due course be given, and the goods can be loaded aboard the ship (or aircraft). The final step is notification of the ship's or aircraft's departure.

Methods of duty payment

> I recommend that any business importing goods regularly opens a deferment account if they can do so. This will aid cash flow and simplify administration.

When goods are imported, you will be required to account for import duty, excise duty if applicable, and VAT. The best way to do this is to open a deferment account. If you are carrying out a clearance using CHIEF, you simply enter your deferment number in box 48 of the electronic form C88, or you can make the equivalent CDS submission, and HMRC will charge VAT and duty to that account. At the end of the calendar month, the amount is totalled and drawn from your bank account (by direct debit) on the 15th of the following month (the excise duty accounting period runs from mid-month to mid-month, with payment near month end). This gives you an average of 30 days' credit on duty and VAT, ie at least 20 per cent of the value of your imports, which is a significant benefit to any business. Statements should be checked thoroughly, as it is not unknown for the wrong deferment number to be used, deliberately or accidentally.

To set up a deferment account you will need a guarantee from your bank or an insurance company to pay on your behalf if you go bankrupt. The guarantee will be required for double the monthly limit of the deferment account, eg £100,000 for a £50,000 limit, as it is in theory possible to fully utilize the account for both previous and current month by the 13th, then enter bankruptcy on the 14th before payment is taken. There will be a charge for providing the guarantee, and banks will be unwilling to do so if the business has financial problems.

Agent's account

Most agents have their own deferment account, and subject to your creditworthiness may be prepared to make a payment on your behalf. They will usually insist upon immediate reimbursement, and make a charge (typically 2–3 per cent) for use of their account.

These arrangements are useful, in that clearance can proceed quickly, but cash flow advantage is lost and the charges will usually exceed those that would have been made by your bank for your own deferment account.

Flexible Accounting System account

A Flexible Accounting System (FAS) account may be the best option for a new business that lacks the creditworthiness to obtain a bank guarantee. In simple terms, you deposit money into the account and must keep it in credit. HMRC then draws from that account to pay duty and VAT on imports. There is no charge for setting up a FAS account, but rather than a cash flow advantage there is a cash flow disadvantage as the money required for payments must be deposited in advance.

Community transit, common transit and TIR

Community transit (CT) is an EU customs procedure that allows goods on which duty has not been paid to move from one point in the EU to another (including from one point to another in an individual member state). Examples might be beer shipped from a bonded warehouse in Poland to a bonded warehouse in Belgium; or goods from the USA that have been airfreighted to France but are trucked to Italy before duties are paid. The procedure also applies to shipments to 'special territories' of the EU (eg the Canary Islands).

A similar procedure, known as common transit, is used for shipments to European Free Trade Association countries, eg Switzerland. One of the first definite announcements about Britain's departure from the EU was that the UK would remain a member of the Common Transit Convention in its own right (Logistics Manager, 2018b).

The *Covention Douanière Relative au Transport International de Marchandises Sous le Couvert des Carnets TIR*, or Customs Convention on the Transport of Goods Under Cover of TIR Carnets (known as TIR), was signed in Geneva in 1975. It is now accepted by countries including China, India, USA and Argentina: Oman was due to join the group on 29 May 2019 to bring the total number of countries to76 (UNECE, 2018). Under the original system, if a journey would pass through one or more countries en route to the destination, a paper carnet was issued for each load. The vehicle was sealed to prevent access during the journey, carried a TIR plate, and the presentation of the carnet to customs when entering and leaving each country would eliminate the need for customs formalities at every border crossing.

The paper control methods for both these procedures have been replaced by the New Computerized Transit System (NCTS). Pre-registration is required before you can use the service. It can be accessed online, or an EDI link can be established for frequent users. A guarantee will need to be provided to customs, in case of failure to comply with requirements, such as a TIR trailer entering but failing to exit a country that it was due to transit, to evade payment of duties.

Customs warehouses: How to delay payment until goods are used

The above section on import clearance assumed that the goods would be cleared into free circulation, with duty and VAT accounted for immediately. However, there is an alternative. A simplified declaration can be made at the port of entry, and the goods taken to an authorized customs warehouse (there are two types – private and public). Goods are then held at the warehouse until they need to be shipped, eg after they have been sold. At this point a further declaration is made to customs, the VAT and duty are charged to a deferment account, and the goods can be shipped. If the goods have been held for a period of several months this will again be a significant aid to cash flow, and whilst there will be an additional administration cost for a customs warehouse, this may be money well spent.

If goods are sold abroad outside of the customs area, they can be re-exported directly from the warehouse with duty or VAT never being paid.

Pre-authorization from HMRC is required before each new customer is permitted to make their first deposit into a public customs warehouse.

A specialized type of warehouse, called an excise warehouse, must be used for the suspension of duty on alcoholic drinks. A number of special restrictions, including enhanced security measures and minimum levels of throughput, apply to such warehouses.

Inward and outward processing relief

Inward processing relief (IPR) and outward processing relief (OPR) are two methods of avoiding customs duty. In simple terms, IPR works on the principle that duty is not paid on goods that are imported, processed, then re-exported within a specified time frame (normally six months). A simple case would be a single item that has been sold abroad and become defective. Under the terms of the warranty given to the customer, it is returned to the manufacturer in the UK, repaired and returned to the customer. If goods are declared as IPR goods on import and export, duty and VAT will not be payable.

One potential pitfall is the use of this procedure for very small-value consignments. I have known companies to use IPR to save a sum of, say, £5 in duty – the cost of administration alone will far exceed this saving.

A more complex example would be the import of air conditioning units that are fitted to cars on a production line, and the completed cars exported. IPR can again be used, but pre-approval by HMRC should be sought, and they will probably make a stipulation as to the minimum percentage of units that must be re-exported.

OPR works on the reverse basis, and can apply to goods that are exported, processed and re-imported. An example would be a suit of clothes which is made in the UK, sent to a non-EU country to have buttons and zips fitted, and returned to the UK. Duty would be payable only on the increased value arising from the buttons and zips. OPR can also be used if goods are being temporarily exported for repair under a warranty.

For either IPR or OPR, a guarantee may be required in case of non-compliance, for example failure to re-export an item within the permitted time. Specialist advice from an experienced agent is recommended if you wish to set up such a scheme, but further details can be found in Annex D to Customs Notice 3001.

Exemption from duty

Generalized system of preference

This is a system designed to encourage trade with the poorest countries of the world by partly or wholly exempting imported goods from those countries from import duty. Donor countries, ie those offering the benefit, are:

- Australia;
- Belarus;
- Canada;
- EU countries;
- Iceland;
- Japan;
- Kazakhstan;
- New Zealand;
- Norway;
- Russia;
- Switzerland;
- Turkey;
- USA.

Some countries, eg China, were former beneficiaries of the scheme but are no longer, due to improved affluence. Some others are excluded due to human rights violations. The list of countries to which the benefit is offered varies by donor country (UNCTAD, 2015).

Other exemptions from duty and VAT

There are a variety of circumstances in which relief from duty and/or VAT can be obtained. Most of these are specialized and will not concern most readers of this book. Here is a list with the relevant customs notice number in brackets so that those who feel they may be able to benefit can obtain further details:

- goods for charities (customs notice 317):
  - basic necessities for the needy, eg food, medicines, clothing, blankets, orthopaedic equipment, crutches;
  - goods donated free from outside the EU to be sold at a charity event;
  - office and other equipment needed to run the charity (motor vehicles other than ambulances are excluded);
  - goods for disaster relief within the EU;
- scientific instruments (340):
  - applies only to public institutions, eg universities and NHS hospitals;
- donated medical equipment (341):
  - applies only to charitable donations to hospitals, etc;
- museum and gallery exhibits (361);
- certain antiques more than 100 years old (362);
- inherited goods (368):
  - any goods included in the estate of the deceased intended for your personal use or for meeting your household needs. Examples are:
    - jewellery, stamp collections, bicycles and private motor vehicles, caravans, trailers, pleasure craft and private aircraft;
    - household furnishings;
    - family pets and horses;
    - portable items (such as doctor's bag, musician's instruments, photographer's cameras and equipment) used by the deceased in their trade or profession;
- goods for disabled people (371);
- commercial samples (372);

- visual and auditory materials (373):
  - must be of educational, scientific or cultural value;
- and finally, decorations and awards (364):
  - if you win a Gold Medal at the Olympics you will not need to pay duty when you bring it home!

Goods intended for sale in duty free shops, for military use, or for use on ships and aircraft are subject to specific and sometimes complex regulations. If you are involved with such matters, you will need specialist advice.

Conclusion

Customs regulations for any country, or grouping such as the EU, are many, varied and complex, with numerous exceptions to the various rules. It is worth reiterating the vital point that it is incredibly unwise to try to defraud or deceive customs authorities in any country – you are likely to suffer severe consequences that far exceed potential gains.

It is, however, possible to use the regulations to your advantage. Payment of duty and VAT can be delayed by use of deferment accounts and customs warehousing. There are also legal methods of avoiding unnecessary payments by use of inward and outward processing relief, and in certain other specific circumstances. I would recommend taking full advantage of these opportunities whenever you can.

Export documentation 17

Logistics is a very broad discipline, and in practice logistics managers are often called upon to handle matters that might not strictly speaking be part of their remit. It is often considered that 'it is up to you to get it there', which may mean involvement in other areas of the business.

Export documentation provides examples of this. Strictly speaking, the preparation of invoices should be carried out by the accounts or sales department, but in my experience if delays in customs in the destination country are to be avoided, the logistics manager will at least need to provide input regarding the correct layout and content for those invoices.

It should be borne in mind that requirements for different countries vary widely. If you are shipping to a wide variety of countries throughout the world, I would recommend subscribing to a manual that provides country-by-country information, either online, or in loose leaf print format (with regular updates provided so that individual pages can be replaced when rules change). These guides are not cheap, but many people find the money is well spent. Options for UK exporters include *Croner's Reference Book for Exporters* and *Tate's Export Guide*.

Certain documents require an original signature from an authorized officer of the company – this can be a director, partner or proprietor, or someone such as a logistics manager who has been authorized in writing to do so.

Chambers of commerce

Responsibility for the issue of many documents required for export, such as certificates of origin, rests with chambers of commerce, both in the UK and elsewhere. They will also be able to offer help and advice for inexperienced exporters. They naturally charge fees for documentation, but rates are reduced for members, and unless you require documents only on a very occasional basis, the savings are likely to pay for the membership fee quite quickly. I would therefore recommend that you seriously consider joining your local chamber.

Most chambers will permit you to register to apply for documentation online, for example using the e-cert system (Chamber International, 2018b). This involves completing forms onscreen rather than in paper format, but the requirements are unchanged. This is a less time-consuming process – you can cut and paste repeat information and do not need to visit the chamber or post items – and I would strongly recommend it.

Commercial invoices

For almost all destinations, a commercial invoice will need to accompany the goods: there are some exceptions, such as movements within the EU Customs Union. At least one copy of the invoice must carry an original signature – the number of copies required varies by country. For example, Australia requires two, of which one may be a copy signature; but Bangladesh requires six, all with original signatures. Printing your name and title under your signature is a good idea.

The invoice is used by your agent for raising documentation relating to export customs clearance; and by import customs authorities to ascertain the nature and value of goods, to ensure that the correct amount of duty is paid. Do not be tempted to under-declare values on an invoice – this is a customs fraud and can attract harsh penalties.

I would always recommend erring on the side of providing too much information rather than too little. For example, not all destination countries require you to state the country of origin of goods on the invoice: however, this may help to avoid delays. Details on the invoice could include:

- invoice number and date;
- seller's name, address and EORI (see Chapter 16):
  - shipper's name and address if different;
- buyer's name and address:
  - destination name and address if different;
  - importer licence number or similar if appropriate to that country;
  - purchase order number;
- numbers and descriptions of all items:
  - if using model numbers follow with a description, eg rather than XYZ model 123 say XYZ model 123 electric fan;
  - for bulk commodities ensure units are clear, eg 1,000 litres;

- price per unit, for each line item, and in total:
  - ensure currency is clear, eg US$ rather than $;
  - include ancillary costs such as freight as separate line if appropriate;
- customs HS tariff code of each item (see Chapter 16);
- country of origin of each item;
- freight terms (see Chapter 15) including place (eg FOB Melbourne rather than simply FOB);
- payment terms (eg within 30 days of delivery);
- packaging details, including weights and dimensions;
- transport details if known, eg ports of departure and arrival, bill of lading or air waybill number, container number, vessel name;
- any specific additional requirements for an individual country (eg Bangladesh requires full name, address, phone, fax and email of the manufacturers of each item).

Many accounting systems lack the functionality to produce an invoice with all the above information. You may therefore need to set up a template on a spreadsheet to produce export invoices that comply with legal requirements, and cross-reference these against invoice numbers generated by the system. Alternatively, they can be generated by the e-cert system (Chamber International, 2018b).

For some countries, such as the United Arab Emirates, invoices must be legalized by the appropriate embassy in the UK before dispatch. You will be able to arrange this through your chamber of commerce. Fees will be payable, and the process will take time: you should take account of both in your planning.

Packing lists

Some countries insist that a packing list is also provided, but in my opinion this is a good idea anyway. The packing list need not be complex – it should include basic details from the invoice such as seller, buyer and invoice number; and contents, marks, weights and dimensions for each package.

It will help customs to carry out an inspection if they wish, and also make checking the goods on arrival easier for your customer, reducing the number of false claims for missing goods.

Certificates of origin

Probably the next most common document required to support exports is a certificate of origin (C of O). Some countries, eg Saudi Arabia, require them for all shipments; others only for certain commodities, eg Malaysia (Croner-I, 2018). They might also be necessary to claim preferential duty rates: for example, goods of New Zealand origin attract reduced duty in Australia, Brunei, Cambodia, Indonesia, Laos, Malaysia, Myanmar, Philippines, Singapore, Thailand and Vietnam (New Zealand Chamber, 2018). Finally, their provision may be a condition of contract or of a letter of credit.

There are two main types in use in the UK as at early 2019 – the Arab–British (see below) and EU. The latter will no doubt be replaced in due course by a similar form that does not bear the name of the EU. Elsewhere, the North American Free Trade Agreement form is used in Mexico, Canada and the USA; other countries including Australia and New Zealand have their own forms, issued by respective chambers of commerce.

Before applying for a C of O for the first time, you will need to complete a declaration, which amongst other things indemnifies the chamber should you provide false information. It must be signed by a director, company secretary, proprietor or partner, and include specimen signatures of anyone authorized to sign documents on the company's behalf.

You will need to provide documentary evidence of the origin of goods if that was outside the UK. This should be an invoice from the original manufacturer or a copy of the original certificate of origin, rather than any document provided by an intermediary in the commercial chain. The C of O must be supported by a commercial invoice, and all information must match.

There is sometimes difficulty in deciding the origin of an item. A useful test is whether the same commodity code (see Chapter 15, page 000) applies before and after an action takes place. For example, if parts are imported from around the world and assembled into a car in Derbyshire, that car will have been manufactured in the UK. If, however, a car is imported from Japan and accessories such as a radio and alloy wheels are fitted before re-export, it remains a car and the country of origin remains Japan. It is also important to differentiate between manufactured goods and 'wholly obtained' goods. The latter are either UK-produced raw materials, such as salt mined in Cheshire; or products derived from them, eg meat from a sheep bred and reared in Snowdonia. There is a box to tick on the rear of the C of O to indicate which of these applies, and you should include details of any manufacturing process. If you are re-exporting imported goods, full names and addresses of manufacturers must be provided on the rear of the application

Paper forms can be purchased from your chamber of commerce. They consist of three parts:

1. certificate (buff coloured with security pattern);
2. control copy (yellow);
3. application (pink) – this will be retained by the chamber.

The form must be typed rather than handwritten; completed in full; and the application copy signed by an authorized signatory, with date and place of signature. Under the last item to be listed, you should draw a line to ensure that no additional items can be added.

Common errors when completing a form include:

- failure to include a country (eg United Kingdom) in the exporter's address;
- poor goods descriptions – brand name or 'spare parts' is not sufficient;
- misunderstanding of 'origin' (see above);
- failure to declare marks and numbers on packages. In theory you can declare them as unmarked, but I would never recommend shipping unmarked boxes as they are very likely to go missing. If only customer address appears you may enter 'fully addressed';
- failure to declare packaging details, ie the number of boxes, drums, reels, sacks, pallets, crates, etc;
- incorrect designation of countries. You should use:
  - United Kingdom rather than UK;
  - European Community – France (for example) for EU countries;
- using imperial rather than metric weights and dimensions;
- failure to declare both gross and net weights;
- using correction fluid or erasing items. If you do make an error you should cross through the incorrect details, write the corrected information beside this, sign the correction, and have the correction stamped by a chamber of commerce.

This section listing common errors is based on Chamber International (2018a).

Provided they are satisfied with the information provided, the chamber will stamp and sign the form. The top copy should them be shipped with the goods.

> It is wise to keep copies of each C of O indefinitely. They may be required if, for example, goods are to be returned or re-exported, which can happen years after the original shipment. Queries can also be raised by customs long after an event.

Arab–British Certificate of Origin

Some countries do not accept the EU C of O and insist on the Arab–British version instead. These countries are:

- Algeria;
- Bahrain;
- Comoros;
- Djibouti;
- Egypt;
- Iraq;
- Jordan;
- Kuwait;
- Lebanon;
- Libya;
- Mauretania;
- Morocco;
- Oman;
- Palestine;
- Qatar;
- Saudi Arabia;
- Somali Democratic Republic;
- Sudan;
- Syria;
- Tunisia;
- UAE;
- Yemen.

In this case the three parts of the form are:

1. original (white);
2. control copy (green);
3. application (blue).

The form must be typed and completed in English or Arabic. Full names and addresses of manufacturers must be listed on all copies, and all supporting documents must be attached. Both green and blue copies must be signed (Croner-I, 2018).

After approval by your local chamber of commerce, it will be forwarded to the Arab–British Chamber of Commerce (ABCC) for their approval. In some cases, it will then be forwarded again to the relevant embassy or consulate for legalization and returned to you via the two chambers. It is possible to ask for documents to be returned to you by the ABCC and for you to forward them to the embassy yourself, but I would not recommend this.

Some countries require an additional declaration – for example Libya insists that Israeli ships and aircraft are not used for transport during the import process (Croner-I, 2018).

Other forms certifying origin

Form EUR1

This form, also issued by chambers of commerce, certifies that goods are of EU origin, and therefore able to claim reduced or zero rates of duty when exported to countries with which the EU has a bilateral trade agreement. There are well over 100 such countries, from Norway to Nauru.

> A key difference to the certificate of origin is that without a valid C of O certain countries will not allow the import to proceed at all; without an EUR1 the import can proceed, but a higher rate of duty will apply.

Form EUR1 can be issued in respect of 'wholly produced' products – in simple terms, those produced by agriculture, fishing or mining within the EU; or those that have undergone 'sufficient transformation'. The latter is a complex question, which may involve consideration of the percentages of EU originating parts, parts originating elsewhere, and EU labour that make up the value of the goods. However, this permitted percentage varies according to the nature of the product!

The best suggestion is to consult HMRC Notice 828: tariff preferences – rules of origin for various countries (UK Government, 2018b). Your chamber of commerce will also be able to offer relevant advice.

It should also be borne in mind that there is a fee for this document: £39.90 plus VAT as at 2018. If your product attracts a duty rate of, say, 2 per cent and a consignment is valued at £1,000, the form will cost twice as much as the duty saving (not considering your own internal administration costs). Care should be taken to ensure that the issue of the form is financially worthwhile.

Turkey

Turkey has a unique relationship with the EU, in that goods in free circulation in either area may be shipped to the other free of duty. Goods in free circulation in the EU are those that have originated in the area, or have been imported and all necessary duties, VAT, and other taxes have been paid. Provided these duties have been paid, the origin of the goods is irrelevant.

This is controlled by Form A.TR, which must be raised by the exporter and endorsed by your chamber of commerce. This form has just two copies – the green top copy travels with the goods, and the white bottom copy is retained by the chamber. A copy of the commercial invoice must be provided when applying, and proof of free circulation such as a copy of the import documentation.

ATA carnets

These can be used to avoid payment of duty on temporary exports. There are three categories in which they may be used:

1. commercial samples;
2. goods for use at fairs, exhibitions and similar events;
3. professional equipment.

They may not be used for goods intended to be sold or hired abroad, goods temporarily exported for repair, equipment for construction of buildings, or certain other categories. When applying for an ATA carnet, you will need to complete a declaration that the goods will be returned to the UK within one year, or a shorter time if stipulated by the destination country (eg six months for India). You will be required to deposit an amount equal to the amount of duty payable should you fail to comply and not return the goods to the UK: this can be cash, bank transfer, bank draft or bank guarantee.

Goods must physically be carried with the holder of the carnet, and not sent separately, for example by post. If the holder is not carrying the goods personally, the representative must either be mentioned on the carnet or carry a letter of authorization.

Once again, chambers of commerce are tasked with issue of ATA carnets, and will offer advice if required. The carnet normally comprises green, yellow and white parts. If, for example, you are a TV camera operator flying from the UK to the USA and back with your camera and associated equipment, you should follow this procedure:

1. At the UK airport, present the carnet to customs. They will:
   - **a** check and certify the green front cover;
   - **b** check, endorse and detach the yellow exportation voucher;
   - **c** complete and stamp the yellow exportation counterfoil;
   - **d** return the carnet to you.
2. On arrival at the US airport, present the carnet to customs. They will:
   - **a** check, endorse and detach the white importation voucher;
   - **b** complete and stamp the white importation counterfoil;
   - **c** return the carnet to you.
3. On departure from the US, present the carnet to customs. They will:
   - **a** check, endorse and detach the white re-exportation voucher;
   - **b** complete and stamp the white re-exportation counterfoil;
   - **c** return the carnet to you.
4. Complete box F on the yellow re-importation counterfoil.
5. On arrival at the UK airport, present the carnet to customs. They will:
   - **a** check, endorse and detach the yellow re-importation voucher;
   - **b** complete and stamp the yellow re-importation counterfoil;
   - **c** return the carnet to you.
6. When you have finished using the carnet, return it to the chamber on or before its expiry date.

If you fail to discharge the carnet correctly, you are liable for duty on the goods: this may be very expensive.

The carnet may also be used as a transit document, for example if someone is travelling from the USA to Russia via the UK, in which case blue vouchers will need to be included in the carnet.

There is a similar system called a *Carnet de Passages en Douane* (CPD) used by some countries for temporary import and export of motor vehicles.

The information in this section is based on Customs Notice 104 – for further details a copy of the notice can be found online (UK Government, 2018c).

Pre-shipment inspection

Numerous countries, mainly in Africa and Asia, insist upon a pre-shipment inspection (PSI) for goods being exported to that country: this may apply to all goods, or only to certain categories. The aim is to ensure that the goods are as described in terms of quantity, quality, value and customs code classification – and that they meet safety regulations where relevant (eg in the case of toys).

An accredited body within the UK such as Bureau Veritas, Cotecna, Intertek or SGS will carry out the PSI, and if all is found to be in order will send a report to the destination country customs authorities.

EAC and GOST R certification

One particularly onerous variation of PSI applies to shipments to the Euro Asian Customs Union, ie:

- Russia;
- Belarus;
- Kazakhstan;
- Armenia;
- Kirghizstan;
- Tajikistan (soon to be included).

EAC certification is required for exports to these countries of a range of products, including personal protection equipment, clothing, toys, perfume and furniture. It can be awarded for a single shipment, a contract, or for a period of one to three years.

GOST R certification applies only to Russia and indicates that a product complies with the laws and safety regulations of the Russian Federation. It is not necessary to obtain both forms of certification for the same products. It is mandatory for some products, and voluntary for others – the latter may be required as a condition of a tender, and it will help to speed up the customs clearance process. It is possible to obtain a certificate of exemption where appropriate.

Comprehensive information must be provided when applying for certification, including description, customs commodity code, technical manuals for electrical goods, composition of garments and pictures of the products. There are additional requirements for some products, for example special sanitary checks for drinks, baby food and personal care products; and calibration checks on measuring equipment.

Obtaining certification can take from two weeks in the case of clothing, to two months or more for equipment to be used in potentially explosive atmospheres.

If you are contemplating exports to Russia or any other EAC country, I recommend that you seek assistance from an experienced organization such as Techsert at the earliest opportunity.

Prohibited goods

It should be remembered that certain countries ban, or have severe restrictions on, the import of certain goods. This applies particularly to Muslim countries, some of which prohibit alcohol, pork-based products, obscene literature and goods related to gambling – including playing cards. Meat may require a Halal certificate, as may cosmetics or medicines that contain animal products.

Other examples may be more surprising, such as Libya's restrictions on envelopes, soap and crystal chandeliers.

> If your company has employees visiting these countries, it is wise to remind them of these restrictions, and to ensure that they do not carry any prohibited items in their luggage. An adult film on a laptop could have very unpleasant repercussions.

Specific product regulations

For a comprehensive listing of all products that may be subject to specific requirements for export to certain countries, you should consult an exporters' guide such as *Croner's Reference Book for Exporters* or *Tate's Export Guide* and/or seek advice from your chamber of commerce. Particular caution should be exercised with respect to:

- Goods with military or paramilitary applications. This includes not just obvious examples such as missiles or firearms but also items such as radar or cryptographic equipment. An export licence for such goods will be required, which will not be granted if you wish to export to certain countries. The Export Control Joint Unit issues licences in the UK.
- Pharmaceuticals. The Medicines and Healthcare Products Regulatory Agency issues licences in the UK where required.
- Waste products: see the Environment Agency website for further details in the UK.
- Live animals.
- Plants and plant-based products.
- Second-hand clothing. This will usually require a fumigation certificate.
- Endangered species and products derived from them – this can include items made from ivory or mahogany. Licences are required under the Convention on International Trade in Endangered Species of Wild Fauna and Flora, which for the most endangered species will only be issued in exceptional circumstances.

Conclusion

In this chapter I have provided practical advice on the documentation most commonly required in relation to exports. If you are not familiar with the various forms, they can appear complex and somewhat daunting. Do not to be afraid to ask for help, as it is far better in such matters to ask those important questions and avoid any possible mistakes.

Miscellaneous supply chain issues

18

In the first chapter I asked the question, 'What is logistics?' I might also have asked, 'What is supply chain management?' These terms can be used synonymously, but at times the meaning of the word logistics can comprise only warehousing and transport.

In practical terms, the remit of the logistics manager within a business is often dictated by the structure of that business. In a small organization it is likely to be much broader than in a multinational, but even in the latter there should be cooperation between departments to develop policies that best serve the organization as a whole.

In this chapter I will look at some of the issues which may or may not fall within that remit. Beginning with inventory planning, I will cover how to set a target inventory level, how to schedule orders to achieve this, and some more advanced techniques for inventory reduction. This chapter will also discuss the final element of the supply chain, ie reverse logistics, and concludes with a subject that is impacted by all aspects of the supply chain – the relative costs of serving different customers.

Inventory planning

A company recruits a new national logistics manager, who is visiting a depot for the first time.

Depot manager: How can I fit more stock into the warehouse?

Logistics manager: Why do you want to fit in yet more stock? You already have 21 months' stock of the most popular items, and 18 months' stock of most of the others.

Depot manager: Well the suppliers like us to buy a lot of their products. And the CEO would not want to take the risk of running out of stock – that would cause a lot of problems. What if one of the suppliers' factories burned down?

Some of the details have been changed in the interests of confidentiality, but the above is essentially what was once said to me when I began a new role.

The fundamental point about inventory is that it has a cost. Not only is there a direct cost in storage, but it also represents a serious drain on cash flow and working capital. Few businesses are cash rich, and the higher the borrowings of a company relative to its underlying financial strength, the higher the cost of further borrowing will become. It is evident that any company, no matter how large, can make better use of large amounts of money than tying it up unnecessarily in stock. Having been involved with multi-million pound stock reductions, this sage piece of advice is just as true in practice as it is in theory. The risk of obsolescence, resulting in write-off of stock, must also be considered.

Having said this, running out of stock will indeed cause problems. You will lose business if you are unable to fulfil orders and will lose customers if this happens consistently. In a manufacturing environment, stopping a production line can be extremely costly. Whilst 'completing' a car without wiper blades and fitting them later is inefficient, the car cannot be completed without an engine, or indeed without engine mounting brackets.

It is recommended that a target inventory level should be treated as exactly that – a target. The target itself should not be treated as a maximum or a minimum.

Setting an inventory target

There is no universal right answer to establishing an optimum target inventory level: it varies according to the nature of the business, and between different items within the business.

Stock levels can be expressed in two main ways. Some businesses measure the stock in terms of days, either physically or in financial terms. For example, if you sell 1,000 items, or £10,000 worth, per day and have 50,000 items or £500,000 in stock, the stock level will be 50 days.

Others use the inverse measure, and express stock levels in terms of stock turns per year: in this case calculated as 365 divided by 50, or 7.3, annual stock turns. If you are importing cheap goods from the Far East to Europe, then several months of inventory may be appropriate; for a major car manufacturing plant several days would be considered excessive.

Another useful measure is the percentage of items that can be met from stock: this is a key performance indicator (KPI) that I would recommend is measured in any stock-keeping environment.

In Chapter 5 I described a Pareto or ABC analysis in the context of warehouse put-away activity, and a similar analysis can be conducted for setting inventory targets, but with one key difference. In many businesses, 20 per cent of line items represent 80 per cent of sales – this is generally known as Group A. In the inventory target context, to this group should be added any product line that may sell in small quantities, but for which the inability to provide that item will have serious consequences. This might be a piece of safety equipment without which work must stop, or a key spare part. For example, manufacturers of combine harvesters sell few replacement pickup reels, but if they were unable to supply one quickly in harvest season the consequences for the farmer might be catastrophic in terms of lost crops. The target inventory for this group should include a safety stock (sometimes known as operational reserve) such that the percentage of orders met from stock will be very close to 100 per cent.

Group B comprises the next 30 per cent of line items, which will probably account for the next 15 per cent of sales. A KPI of, say, 99 per cent of orders met from stock may be acceptable here.

Group C comprises the remaining 50 per cent of items. A KPI of 90 per cent might be appropriate for this group, but it is worth remembering that some items will sell only if in stock: a customer may need to contact a number of potential sources before finding one with the item in stock. This situation occurs in the car spares market, in relation to older models for which parts are no longer kept by retailers, and a stock of one or two may be sufficient for the wholesaler.

I would also add a Group D, consisting of items that fall outside the normal pattern. This might include promotional items, perhaps sold on a WIGIG basis ('when it's gone it's gone'); or stock built up over a period to meet a seasonal peak demand – Christmas crackers would be an extreme example. There may also be items that are rarely used, and may even be bespoke to particular projects, which need to be sourced only on demand: a 25 metre radio mast is an example I personally recall.

In deciding the appropriate level of safety stock, you will need to take into account lead-times, as this will be the key factor in replenishing stocks if they run low. This should not only include time in transit from the supplier, but in relevant cases the time needed for the supplier to manufacture the goods, and/or time for your manufacturing or rework processes. This may not be constant across your range. For example, if you are carrying out security system installations, the parts needed for a particular project might include control units sourced from the USA, with six week lead-times; and a battery (for back-up in case of power cuts) available for next day delivery from a range of local distributors. Clearly the level of safety stock for the former should be higher than for the latter.

Finally, you should consider any contingency arrangements that may be available. For example, goods might be airfreighted rather than seafreighted, and potential alternative suppliers identified in case of catastrophic events such as a factory fire.

With all this in mind, the contrary consideration is what the business can, or should choose to, afford. In a fledgling (or financially struggling) organization, there will be a finite limit to the cash available, the limits of prudent borrowing having been reached. Sometimes a board-level decision will be required, as to whether available working capital is best spent on stock, or for example on new machinery or employing additional staff. I do not recommend ordering stock that you cannot afford to pay for, as this will probably result in credit facilities being withdrawn and the situation worsening. In such situations, and indeed if there is a realistic but finite budget, being able to balance stocks within the funds available is a valuable but often under-rated skill.

Demand planning and order scheduling

Having established the target for stock level in terms of days of stock (or other measure), the next stage is to establish what quantities equate to that target. For example, if your target is 60 working days of stock, and you sell 100 of item X per working day, then the average stock will need to be 6,000. However, in order to reach this point, you will need to establish what the probable demand will be.

Software is available to assist the process, either as a stand-alone package or as part of a broader enterprise resource planning or materials requirements planning/manufacturing resource planning system, but in small businesses it may be necessary to utilize a spreadsheet to analyse downloaded data.

The main elements in the process are as follows:

1 Data analysis. This will usually be driven by sales, and good communication with the sales and marketing departments is therefore vital. A realistic estimate needs to be made of what customers will buy, and when. The process needs to include:

   a Past sales performance. In most companies the best indicator of future sales will be past sales. Generally, this should be the benchmark for analysis, and should always be used as a sense check to ensure that projections are realistic. In a mature and stable environment there may be little change year on year. In a new business, launch predictions can be much more difficult, and must be based on market research and projections.

   b Customer provided data. If you are selling components to a manufacturer, they should be able to provide an estimate of their requirements, although they are unlikely to make a firm commitment. Retailers and distributors may also do so. Such figures, if available, will be very useful.

   c Long-term trends. Is your product in a growing or declining market? Biotechnology and virtual reality are growing rapidly. Boating and private flying as leisure activities have declined since their peak. This will be reflected in future sales.

   d Market share. Many businesses target their sales force with increasing market share. In many industries, every sales director will have an objective to do so. Clearly not all can succeed, and you therefore need to challenge assumptions to ensure that they are realistic and not driven solely by over-optimism.

  e Seasonality. It may be necessary to build up stocks in advance of a peak season: early summer months for air conditioning units and clothes pegs; August for school stationery; mid December for cranberry sauce and carving knives (some families now use a carving knife only once per year for the Christmas Day turkey); less predictable peaks for waterproof clothing.

  f Promotions and new product launches – and the opposite. If a promotion or new product launch is planned, then clearly stock needs to be on hand ready for that promotion. This must include secondary sales – if a new model of car is launched, car dealers will need an initial stock of lights for the aftermarket. However, the reverse must also be considered: if a new model is launched, the existing model will become unattractive, and only be able to be sold at a discount or not at all. Stock of old models should be run down progressively to avoid this.

  g Competitor activity. If you know your competitor will launch a new product, it is better to be realistic and accept that this will reduce your sales, at least temporarily.

  h Predictable events. Sales of beer will rise during international football tournaments, for example.

  i Unpredictable events. There will be a surge in demand for replacement car batteries after a severe frost, and for particular cooking ingredients after a recommendation from a TV chef.

2 Forecasting demand. Taking past sales as a base, and adjusting for whichever of the above factors may be appropriate, it will be possible to establish your projected requirements. For example, sales of cameras have declined in recent years, especially at the cheaper end of the market, as people use their mobile phones to take pictures. However, there remain seasonal peaks in the summer and approaching Christmas, and other factors such as promotions will also have an effect. You might start the analysis by determining previous years' sales, including the seasonal peaks, and adjust for a probable continued decline. Any other relevant elements can then be taken into account, to arrive at the best available projection. Please also note:

  a If you are manufacturing, a further step will be necessary, to convert the projected sales of a completed item into a projected requirement for its components.

  b Ensure that you include ancillary items, such as packaging materials, instruction leaflets, or batteries in your projection.

3 Stock target calculation. This is the mathematical process described above, combining projected usage and target stock level in terms of days of stock, to establish a figure for the number of items required. I would recommend that this exercise be conducted annually, and the projection shared with your suppliers to enable them to plan their production and/or purchasing accordingly.

4 Revision of analysis. The projection needs to be revisited regularly. In some environments, such as automotive manufacturing, teams of analysts monitor stock levels in real time and adjust orders throughout the working day. In a small organization a weekly review will probably be sufficient.

Order scheduling

Having completed the above analysis, it is now possible to schedule and place orders. For the purposes of this discussion I have assumed that suppliers have been selected and prices agreed, and this is a call-off order specifying quantity and delivery date.

This is again a mathematical process, in which current stock, target stock, usage, lead-times and goods in transit must be taken into account. It is best illustrated by an example. Item Y had projected sales of 100 per week throughout the year, and a target stock level of 800. It has sold better than expected, 150 per week, as we gained new customers due to a competitor's quality problems. Stock is currently at 500, with another 400 expected in two weeks. I need to place an order for delivery in four weeks, when I will be looking to increase stock to account for higher sales. As I want my average stock to be 1,200, I will aim for 1,350 immediately after the delivery, to fall to 1,050 in six weeks when the following order will be due. I can therefore calculate the size of the order as 1,350 (target) + 600 (expected usage before order arrives) – 500 (current stock) – 400 (stock in transit) = 1,050.

The same principles apply to perishable goods, but scheduling must be more agile to account for short and limited shelf-life.

Conducting these analyses can be time-consuming, but it is necessary. Policies such as ordering the same quantities every week quickly create imbalanced stocks as usage is never completely constant, and I would strongly recommend that they are avoided.

> In some environments, you will be offered cheap goods. It can be tempting to accept such offers, and if you will use the goods in a reasonable time and are satisfied as to their quality that is all well and good. However, one of the first principles of purchasing is that if you do not actually need an item, it does not matter how cheap it is, it is *not* a bargain.

Inventory reduction

There are a number of supply chain methodologies that can be used to reduce the cost of inventory – some of these will be described later.

Much can be achieved by basic housekeeping, and the elimination of waste. This will especially be the case if you are moving into a new position and inheriting an imbalanced inventory. Examples include:

- Centralize order placement. There are countless examples of one person ordering items that are over-stocked elsewhere in the company. This should be avoided if orders are routed through a central coordinator, who at least analyses and if appropriate challenges orders, or better still develops and places the orders themselves. However, this will not be effective if the coordinator merely acts as a post-box.
- Transfer rather than purchasing goods. If one depot is over-stocked with an item but another depot under-stocked, an internal transfer will be the best option. This is one advantage of centralized order processing.
- Avoid equal orders across a range of products. For example, you may have a team of engineers who require safety boots. An initial order is placed for 12 pairs in each size. After a time, all size 10s have been issued, so another order is placed for 12 pairs in each size. The process is repeated, resulting in a large stock of unused size 6s. Clearly order quantities should be adjusted in line with usage of each individual item.
- Consider whether slow-moving stock can be used as an alternative for faster moving stock. As an example, let us assume that a 17-element TV aerial costs 25 cents more than a 15-element. The former is necessary in about 0.1 per cent of installations, but there are enough in stock to last for 20 years at current usage. However, it can be used in place of a 15-element at any time. It is therefore sensible to use most of this stock before buying further 15-element aerials.
- Manage obsolescent and unwanted stock. In the ideal world there would be no such stock, as inventory levels would have been progressively run down in advance. However, that is very rarely achieved in the real world. If you have stock you do not need, it is wise to act quickly – the longer you wait, the more difficult the task will become, and the more the stock will devalue. The options you might consider include:
  - Rework or refurbish to a newer specification. This will be possible only in a small minority of cases but is the first option to consider.
  - Use the stock slowly. There may still be some demand for older models, and it may be necessary to sell off remaining stock gradually over a couple of years.
  - Similarly, out-of-season goods can be sold in small quantities, or retained until the following season (eg some people may give a lawn mower as a Christmas or wedding gift; there will be another major selling opportunity the following summer).
  - Parts originally intended as production components can sometimes be sold in the spares market. This is not always the case – for example, complete car exhaust systems are supplied for production; separate sections are sold in the aftermarket. Repackaging will also be necessary.

- Cannibalize the items for spare parts. This is another option that will be available only in rare circumstances, but opportunities to do so will be welcome.
- Sell it through an alternative channel:
  - A retailer may sell unwanted lines in a sale or a cut-price promotion, or you may be able to convince your customer to take a bulk order at a reduced price. This is often the best approach with goods approaching a 'best before' or 'use by' date.
  - At least one airline sells goods no longer needed for in-flight retail (following range changes) through its staff shop.
  - Some companies specialise in buying up job lots of cheap goods for sale to developing nations – a lot of clothes donated to charity shops are sold in Sub-Saharan Africa.
  - If all else fails you could try selling on eBay. I know of one company that employs someone full time doing precisely this.
- Return to supplier. Some organizations, such as TV shopping channels, include a sale or return clause in their contracts. You might be able to do so, but the costs of offering this service will obviously be taken into account in the supplier's selling price to you. Without such a clause, the supplier may accept returned goods, but will almost certainly make a handling charge. If the products are of no more use to the supplier than to you, then they are unlikely to do so.
- Sell for scrap or low-value use. For example:
  - I once inherited a stock of electric cable that could no longer be legally used as it did not meet 'low smoke and fume' fire regulations. The only viable alternative was to sell it to a metal recycling company as scrap copper.
  - Food no longer fit for human consumption may be saleable as animal feed.
  - It is possible by following correct procedures to reclaim excise duty on alcoholic drinks that are destroyed.
- Secure destruction. If, for example, you discover a safety fault with a batch of electrical goods, you may decide that it is not viable to repair them, so they must be disposed of. However, if you simply place them in a skip with the intention of being sent to landfill, they may find their way onto the market, and be sold to unwitting members of the public, harming your reputation. Major waste disposal and recycling companies such as Veolia offer secure destruction services, ensuring that this cannot happen. You can send a representative to personally witness the destruction, or be supplied with a DVD (Veolia, 2019).
- Donation to charity. Small local charities may be pleased to accept your donations, and put them to good use. It is wise to double-check that your donated items are used as intended.

- Landfill. This should be a last resort – you will incur disposal charges and landfill taxes, and there will be a negative environmental impact for no benefit. It may, however, be unavoidable.

Advanced inventory reduction strategies

In addition to exercising good disciplines and rigorous analysis as described above, there are strategic policies that can be implemented. The decision to do so should only be taken after careful consideration and planning – to attempt this without the necessary control procedures in place can have disastrous consequences.

Just-in-time

Just-in-time (JIT) had its origins in the Japanese automotive industry, and was initially driven by road traffic laws at the time, which prohibited large goods vehicles on the country's roads. However, it soon became clear that there were other advantages, such as eliminating large quantities of stock at the side of production lines, and it has since been adopted by automotive and other manufacturers worldwide.

The basic premise is that parts are only delivered just before they are needed. Most suppliers will make daily deliveries, although there may even be hourly deliveries of some key components.

To operate effectively, everything must proceed according to plan. In practice therefore, some reserve stock needs to be held in plant or nearby. If goods are being seafreighted from another continent, a buffer stock in a local warehouse is almost essential. Traffic disruption, quality problems with deliveries, and a need to change the planned production mix (known as re-segmentation) can all create problems, and contingency planning must be in place to cope with such eventualities. I recall that as a graduate trainee, I as the most junior member of staff was dispatched to the parts depot to collect some parts to prevent a line shortage. However, methodologies have improved in the years since. With careful preparation JIT does work, and it is an essential element to competitiveness in some industries.

Sequencing

The principle can be taken further by sequencing goods. For example, a motor manufacturer fitted bumpers to match car body colour. This used to be achieved by keeping stillages of the various colours of bumper near to the production line, but these occupied a lot of space, which slowed production as an operative needed to walk up to 10 metres to retrieve the correct bumper. I therefore developed a service whereby the stillages were delivered to a warehouse about 1 km from the plant. Production schedules were passed to the warehouse, and if for example the first three cars were to be red, blue and white, then bumpers would be arranged in a stillage in that order. Deliveries to plant were made every two hours, and production line efficiency was substantially improved.

In the retail environment, a similar principle can be applied to shelf replenishment. Roll cages can be loaded so that as an operative walks down an aisle the package at the top of the stack will be the one to be placed at the near end, the bottom package at the far end. It is often said in the retail industry that the last 50 metres of the supply chain are the most difficult and this is an example of how efficiency can be improved.

Buy or produce to order

This is another variation on a similar theme to JIT. The principle is that you only order goods when you know that you have made a sale. In some circumstances this is the most logical approach. To use an extreme example, no one would build a multi-million dollar satellite unless it was part of a fully funded programme for launch. Similarly, whilst car manufacturers build standard specification cars for stock, as they can be confident of a sale, a top of the range vehicle with an unusual mix of optional extras is only likely to be built as a 'sold order'. Bespoke furniture and window frames are other examples. Similarly, whilst a small decorating company might maintain a stock of white paint, they would be unlikely to buy a particular pattern of wallpaper unless a customer had requested it.

This approach may also be appropriate as a product approaches the end of its life, or if you sell a particular product only to one customer.

Buying to order was once common practice in mail order – hence newspaper advertisements often said 'Allow 28 days for delivery' – but this is not attractive in the current marketplace. It is specifically forbidden in some circumstances – for example UK media regulators forbid TV shopping channels from selling any goods that are not already in their warehouse.

Consignment stock

Consignment stock is owned by one party, but another party keeps it on their premises and attempts to sell it. The owner only receives payment if and when the goods are sold. This differs from a sale or return arrangement, where the goods are paid for, but a refund is then made for any unsold goods returned to the vendor.

For the seller, there are obvious cash flow advantages – they have a chance to make a profit by selling goods, without expending capital by buying stock or running the risk of being left with unsold stock. There will, however, be costs in marketing the goods, whether by display in a retail shop, advertising online, or other means; you will be expected to take responsibility for lost or damaged stock; and profit margins are likely to be lower than in a conventional sale.

For the vendor, this can be a method of trialling a product in a new market without the expense of setting up a major sales infrastructure. It might be a way of determining whether a particular product range will sell in a particular area, or of accessing markets that you would not otherwise be able to access. There is a risk that the other party will sell the goods and not pass on the payment, whether due to bankruptcy or dishonesty, and it is wise to ensure the reliability of the other party before entering into an agreement.

The system can also be used in a manufacturing environment. A supplier delivers components to the manufacturer, who retains them until they are needed, and pays only when they are used in production.

Virtual warehousing

The concept of a virtual warehouse is that goods which superficially appear to be held in one warehouse are in fact held in a number of different locations, but are equally available for use or purchase. The locations might be a number of different warehouses operated by the same company, or even supplier premises.

For example, if a customer in the UK orders three items from a retailer's website, one may be shipped from the retailer's UK depot; the second from their depot in Germany; and the third directly from a supplier in Italy.

I have also heard this system referred to as fictitious warehousing. However, statements such as 'your goods will be held in a fictitious warehouse' would be very confusing to most people, and I recommend that the expression be avoided.

To operate virtual warehousing effectively requires complex and robust infrastructure and management, and I would not advise doing so without very careful planning.

Reverse logistics

The term reverse logistics has been in use only since the 1990s, but the concept has been around for much longer. In short, it refers to movements in the opposite direction to the original (or forward) logistics system, ie returns from your customers back to you, and/or from you back to your suppliers; and associated activities.

With the growth of ecommerce and mail order, the importance of reverse logistics has increased. In the UK, the basic law for face-to-face sales (eg in a shop) is that refunds need only be offered if goods are not as described, or faulty. However, for distance selling (internet, mail order or telephone sales) there is an automatic 14-day period in which an order can be cancelled and a refund demanded (UK Government, 2019). Some retailers now offer a similar service even for sales in their shops, and some UK TV shopping channels offer a 'no questions asked' returns service to their customers – and all goods returned under the scheme are sent back to the supplier. If you are supplying such an organization, it will be your problem to deal with these returns.

In the past, returns were a low priority in many companies, and were often left to be dealt with 'some time when we are not busy' – a time that never came. A modern business cannot usually afford to take such an approach.

Some customers make full use of the returns process. They may, for example, order three garments, their own size plus the one above and one below, with the intention of returning two. Top-of-the-range men's suits returned 'unworn' but with a wedding invitation in the pocket are not unknown. Similar issues also occur in the commercial sector. My personal favourite example is a car dealer who claimed a car bonnet was damaged on receipt: on examination, it was found that they had painted the panel, and chipped the paint.

It is wise to establish a returns procedure, and agree this with your customer. Ideally, it should include the following:

- Time limit for notification. Commercial and legal requirements may limit room for manoeuvre in this area, but if you are able to do so, specify as short a time as possible, to prevent claims for goods missing or damaged on arrival, that have actually arisen from incidents within your customer's operation.
- Authorization. I recommend that a returned material authorization (RMA) number be allocated to each return, to be marked on the package. You should monitor these, and perform a statistical analysis to identify recurring issues. If a particular customer has a larger than expected share of issues, this may indicate a problem – they may be abusing the system; they may have poor checking procedures; or there may be a careless delivery driver. Recurrent returns of a particular item may suggest a problem with your manufacturing process or a supplier.
- Specified carrier. Ideally, you will be able to collect goods on your next regular delivery. If not, specify the carrier to be used, and if possible produce a label online that the customer must attach to the package. This will avoid poorly labelled packages that never reach you (a surprising number of customers put their own address on the outside of return packages), or which cannot be cross-referenced to an RMA number. Returned packages appearing from numerous different sources create unnecessary problems.
- Inspection on arrival. This should confirm that there is a valid RMA, and that the goods returned match the details of that RMA. You may find that you have been

returned a competitor's product, or one which bears no resemblance to the item: a kitchen sink received when we expected a case of wine is perhaps the most extreme example I have witnessed.

- Classification. A more detailed examination of the item can be conducted later. The details will vary, but if, for example, you are processing electrical goods returned by a shopping channel the classification process might include:
  - Boxes still on original pallet, and never sent to customer. Return to stock for sale.
  - Outer packaging never opened. Remove or cover address label. Return to stock.
  - Packaging opened, but goods in mint condition and appear unused. Re-box, replace instruction leaflet if missing, return to stock.
  - Goods appear to have been used. Remove from box, plug into electric supply to test.
    - If not working, return to your supplier; or if this is not possible, remove any detachable parts, then set aside for secure destruction.
    - If working but marked, clean, repack and put into stock for sale as used item
    - If working but damaged: replace any damaged detachable parts with those removed from non-working units; repackage; sell as used; if non-replaceable parts damaged, set aside for secure destruction.

Obviously, the ideal solution for any returned goods is to resell them at full price, but if this is not possible the various options for obsolescent or unwanted stock already described can be considered.

You should also have a robust system in place for monitoring the quality of goods you receive from your suppliers, and agreed procedures for their return if necessary.

The returns policy agreed with your customers, and indeed with your suppliers, will impact on the profitability of any contract. This impact should not be ignored.

Cost to serve different customers

This final topic brings together elements of much of what has been discussed previously.

I have on numerous occasions been asked to quote an 'ad valorem' rate for delivery costs, ie one based on a percentage of the cost of the item, regardless of its size, weight, destination or other details. Unfortunately, there is little correlation between freight cost and item value, so this is not a realistic option.

It is recommended that you estimate logistics costs as accurately as possible before finalizing your selling price. This can be done by adding the various elements described in earlier chapters – maybe seafreight, handling in port, customs clearance charge, customs duty, warehouse handling, warehouse storage for an estimated average time and delivery. There is a reality that selling prices may be dictated by what a customer will accept, and market rates relative to competitors. However, you should at the very least be aware of the impact of logistics costs on your margin.

Chapter 7 discussed varying customer requirements. These can have a significant impact on your costs. Let us consider a hypothetical example. The product is an electrical appliance, which is imported from the Far East by seafreight and held in a warehouse in the English Midlands. Costs to this point are ignored for the purposes of this exercise as they are constant. Goods are palletized, 180 boxes to a full pallet, 26 pallets per full artic load.

Let us consider just two hypothetical customers:

- Customer 1. Accepts full artic loads of complete pallets, delivered to the customer's distribution centre near Doncaster:
  - picking and dispatch cost £3 per pallet, or £78 per load;
  - delivery £250 per load;
  - total cost £328 per load, or 7p per carton.
- Customer 2. Requires palletized deliveries to stores, mixed items on pallets, with different quantities for each store. Average 100 cartons per pallet. Each carton must have special label:
  - labelling cost 3p per carton;
  - picking and palletization 20p per carton;
  - dispatch £1.50 per pallet, equals 1.5p per carton;
  - delivery average £50 per pallet, equals 50p per carton;
  - total cost 74.5p per carton.

Clearly every case will be different, but a 67.5p per unit difference in delivery cost due to differing requirements is quite possible. This could represent a significant proportion of your profit margin, and should not be ignored.

Summary

To summarise, the main points covered in this chapter include:

- Establish a target for the best inventory level, with the best balance between:
  - ensuring that customer demand is met;
  - not incurring excessive costs due to additional storage or utilizing too much working capital.

- Schedule your orders such that stock remains as close as possible to that target, by:
  - analysing available sales data;
  - forecasting future demand;
  - calculating correct order schedules to meet that demand;
  - reviewing the calculations regularly.
- Reduce inventory levels where possible, for example by:
  - centralizing order placement;
  - utilizing stock from other depots, or using alternatives;
  - managing obsolete or unwanted stock.
- If appropriate, use more advanced strategies, eg:
  - just-in-time;
  - buy or produce to order;
  - consignment stock;
  - virtual warehousing.
- Manage reverse logistics to minimize wastage.
- Be aware of the variation of costs due to differing customer demands, and take this into account when setting selling prices to those customers.

Conclusion

Some of the topics discussed in earlier chapters impact directly on the remit of a logistics manager, but not directly on the rest of the business. The subjects discussed in this chapter, however, may not in some cases fall within their direct remit, but in either event will impact on both the logistics manager and other parts of the business.

The setting of inventory targets, and maintaining inventory at the right level, will make a significant contribution to the cash flow, and hence borrowing costs and profitability, of the whole company.

The cost to serve different customers, including returns policies, can impact severely on profit margins. It may be the case that your sales team is unaware of such differences, and will happily agree to additional actions without considering the cost implications. I have stressed on several occasions the importance of communication within a business, and this is a clear example of that principle.

AFTERWORD

So, what is logistics?

I started this book by asking the question 'What is logistics?' I quoted the tongue-in-cheek remark that all logisticians do is 'run sheds and lorries' and quoted business guru Tom Peters, who clearly understands the importance of logistics in ensuring that the right things get to the right place at the right time.

Operating warehouses and trucks does form part of the process, but there is a great deal more to logistics than that. You need to choose the right equipment for your warehouse, choose the right trucks and work out how to operate them to their best advantage, maintaining the optimum inventory level. You will probably need to use other modes of transport, whether parcels sent by air or delivered by van, or containers or bulk consignments dispatched by rail or sea. You need to package the goods so that they arrive safely, meet your customers' delivery requirements, and complete the necessary delivery, export and import documentation.

To my mind, this variety is one of the great attractions of working in logistics. I am approaching the final stages of a career spanning more than 40 years and have thoroughly enjoyed it from beginning to end. I hope not only that you have found this book useful, but that you have enjoyed reading it. I also hope that wherever you are in your career, and however large or small a part it plays in your current and future roles, you share my genuine enthusiasm for all aspects of logistics, and that you gain as much enjoyment from your career as I have from mine.

REFERENCES

Almyta (2018) http://almyta.com/abc_inventory_software.asp [Accessed 23/07/2018]
Amazon (2018) https://images-na.ssl-images-amazon.com/images/G/01/magicarp/common/Carrier_Central_Manual_GBAmazon.pdf [Accessed 23/07/2018]
Armstrong Logistics (2018) https://www.armstrong-logistics.co.uk/materials/rha_conditions_of_storage_.pdf [Accessed 18/12/2018]
ARRC (2018) http://www.cisnav.com/index.php?option=com_content&view=article&id=66&Itemid=91 [Accessed 29/11/2018]
ATS (2018) https://www.atseuromaster.co.uk/consumer/tyres/type/budget-tyres [Accessed 16/10/2018]
Australian Government (2018) https://bicon.agriculture.gov.au/BiconWeb4.0/ImportConditions/Conditions?EvaluatableElementId=296783&Path= UNDEFINED&UserContext=External&EvaluationStateId=4e56c64c-3c17-4791-bd91-66e1e929fec4&CaseElementPk=859698&EvaluationPhase=ImportDefinition&HasAlerts=False&HasChangeNotices=False&IsAEP=False [Accessed 09/08/2018]
Axiom (2018) http://www.axiomgb.com/_pdf/Final%20online%20Axiom%20Unipart%20case%20study%20-%208%20May%202013.pdf [Accessed 06/08/2018]
BBC (2017) https://www.bbc.co.uk/news/uk-england-devon-38657674 [Accessed 29/11/2018]
Belt Railway (2019) http://www2.beltrailway.com/ [Accessed 19/1/2019]
Bendi (2016) http://www.translift-bendi.co.uk/downloads/2016-Printable-Landoll-Bendi-B40VAC.pdf [Accessed 02/07/2018]
Brake (2018a) Essential guide to fleet safety for SMEs and EMPLOYERS starting out in road risk management, Brake, Huddersfield
Brake (2018b) http://www.brakepro.org/assets/docs/practitioner-tools/Brakepro-driveradvice-15-drinkdriving.pdf [Accessed 19/10/2018]
Business Insider (2018) http://uk.businessinsider.com/best-selling-trucks-in-america-2018-3?r=US&IR=T/#11-gmc-canyon-32106-sold-in-2017-down-14-over-2016-1 [Accessed 22/08/2018]
BVRLA (2018) http://www.bvrla.co.uk/advice/guidance/returning-your-leased-vehicle [Accessed 11/10/2018]
Canal Plan (2018) https://canalplan.eu/cgi-bin/canal.cgi [Accessed 21/11/2018]
Chamber International (2018a) https://www.chamber-international.com/uploads/files/7-common-errors-on-a-certificate-of-origin.pdf [Accessed 11/12/2018]
Chamber International (2018b) https://www.chamber-international.com/uploads/files/e-cert.pdf [Accessed 14/12/2018]
Chamber International (2018c) https://www.chamber-international.com/uploads/files/how-to-work-with-letters-of-credit.pdf [Accessed 17/12/2018]
CHEP (2018) https://www.chep.com/uk/en/consumer-goods [Accessed 16/08/2018]

Commercial Motor (2018) https://www.commercialmotor.com/news/buyers-guide/truck-finance [Accessed 11/10/2018]
Croner-I (2017) https://app.croneri.co.uk/feature-articles/drivers-hours-and-rest-periods [Accessed 26/10/2018]
Croner-I (2018) https://app.croneri.co.uk [Accessed 10/12/2018]
Dargan, S (2018) *Logistics Manager Magazine*, September, p 33
Davies Turner (2018) https://www.daviesturner.com/ebrochures/Road-brochure/ [Accessed 02/11/2018]
Department for Transport (2009) https://assets.publishing.service.gov.uk/government/uploads/system/uploads/attachment_data/file/409165/ Information_Sheet_Overhanging_loads.pdf [Accessed 29/08/2018]
Department for Transport (2010) Freight best practice: Fuel saving tips, Pocket Guide, HMSO
Department for Transport (2015) https://assets.publishing.service.gov.uk/government/uploads/system/uploads/attachment_data/file/643021/MSRS_Guide_2015_16.pdf [Accessed 16/11/2018]
Department for Transport (2017) https://assets.publishing.service.gov.uk/government/uploads/system/uploads/attachment_data/file/664323/tsgb-2017-print-ready-version.pdf [Accessed 22/11/2018]
Descartes (2018) https://www.descartes.com/pixi-ecommerce-wms [Accessed 25/07/2018]
DHL (2018) http://www.dhl.com/content/dam/downloads/g0/express/shipping/packaging/dhl_express_large_palletised_packing_guide_en.pdf [Accessed 09/08/2018]
DVLA (2018) https://www.gov.uk/apply-for-a-digital-tachograph-driver-smart-card [Accessed 29/10/2018]
Economist (2013) https://www.economist.com/the-economist-explains/2013/05/21/why-have-containers-boosted-trade-so-much [Accessed 29/11/2018]
Emblem, A and Emblem, H (2012) *Packaging Technology Fundamentals Materials and Processes*, Woodhead Publishing, Cambridge
Engie (2017) https://www.engie.com/wp-content/uploads/2017/06/regular_bunkering_op_pr_def.pdf [Accessed 03/12/18]
European Commission (2015) https://ec.europa.eu/transport/sites/transport/files/2015-07-swl-final-report.pdf [Accessed 14/11/2018]
European Court of Auditors (2015) *Inland Waterways in Europe: No Significant improvement in modal share and navigability conditions since 2001*. Non-print legal deposit e-book retrieved from www.bl.uk
Eurotunnel (2018) https://www.eurotunnelfreight.com/uk/about/ [Accessed 14/11/2018]
Fairplay (2016) https://fairplay.ihs.com/ship-construction/article/4276096/first-dual-fuel-vessel-enters-car-carrier-sector [Accessed 03/12/2018]
Ford (2018) https://www.ford.co.uk/vans-and-pickups [Accessed 21/08/2018]
Freight by Water (2018) http://freightbywater.fta.co.uk/export/shared/downloads/modeshiftcentre/case_studies/cory.pdf [Accessed 21/11/2018]
Freight on Rail (2018) http://www.freightonrail.org.uk/FactsFigures.htm [Accessed 12/11/2018]
FTA (2018) *Logistics Report 2018*, Freight Transport Association, Tunbridge Wells

Gefco (1996) *Petit Dictionnaire du Transport*, Transport Gefco, Aire (In French)
Glover, J (2013) *Principles of Railway Operation*, Ian Allan Publishing, Hersham, Surrey.
Guardian (2015) https://www.theguardian.com/technology/2015/jun/15/i-read-all-the-small-print-on-the-internet [Accessed 16/12/2018]
GWR (2015) https://www.gwr.com/about-us/media-centre/news/2015/november/live-lobsters-take-the-train-to-london [Accessed 12/11/2018]
Hallett Silbermann (2018) Excerpts from *Drivers' Handbook*
Herchenbach (2018) https://www.herchenbach.co.uk/temporary-buildings/prices.htm [Accessed 17/07/2018]
HMM (2018) http://www.hmm21.com/data_files/jsp/eng/service/bulk/vessel/PJT_20110421.pdf [Accessed 06/12/2018]
HMRC (2016) https://www.gov.uk/guidance/fuel-duty#rates [Accessed 23/07/2018]
HMRC (2018a) https://www.gov.uk/guidance/fuel-duty#rates [Accessed 03/07/2018]
HMRC (2018b) http://www.uk-customs-tariff.com/Login.aspx?ReturnUrl=%2f [Accessed 19/12/2018]
IMF (2018) https://www.imf.org/external/np/fin/data/rms_five.aspx [Accessed 17/12/2018]
IMO (2018a) http://www.imo.org/en/MediaCentre/HotTopics/Pages/Sulphur-2020.aspx [Accessed 24/10/2018]
IMO (2018b) http://www.imo.org/en/MediaCentre/HotTopics/GHG/Documents/2020%20sulphur%20limit%20FAQ%202018.pdf [Accessed 24/10/2018]
International Railway Journal (2018a) https://www.railjournal.com/in_depth/china-europe-rail-freight-in-it-for-the-long-haul [Accessed 15/11/2018]
International Railway Journal (2018b) https://www.railjournal.com/freight/italian-high-speed-rail-freight-service-ready-for-launch/ [Accessed 22/11/2018]
Internet Retailing (2018) https://internetretailing.net/themes/themes/mamps-and-new-report-show-rfid-increases-retail-sales-by-up-to-55-16097 [Accessed 26/07/2018]
Jungheinrich (2018) https://www.jungheinrich.co.uk/resource/blob/12754/7fc9cd5a328ef867be10f086b0fa4013/data-sheet-dfg-tfg-316-320-pdf-data.pdf [Accessed 28/06/2018]
Kahn-Freund, O (1965) *The Law of Carriage by Inland Transport*, Stevens and Sons, London
Keller, SB and Keller, BC (2014) *Definitive Guide to Warehousing*, Pearson, New Jersey
Kite Packaging (2018) https://www.kitepackaging.co.uk/scp/polythene-tubing/lay-flat-polythene-tubing ≈06/08/2018]
Kurkela, MS (2008) *Letters of Credit and Bank Guarantees Under International Trade Law*, Oxford University Press, Oxford
Leidos (2017) https://www.leidos-supply.uk/sites/default/files/Leidos%20Supplier%20Manual%20Version%202%20September%202017_0.pdf [Accessed 16/08/2018]
Loading Systems (2018) Dock equipment brochure, page 15, https://www.loading-systems.com/en-gb/products/dock-levellers/trucks-and-vans [Accessed 26/06/2018]
Logistics Manager (2018a) https://www.logisticsmanager.com/ford-targets-grocery-deliverers-with-new-vehicle/?utm_source=Subs_LM_Daily&utm_campaign=401150059c-EMAIL_CAMPAIGN_2017_08_01_COPY_01&utm_medium=email&utm_term=0_27b19fa38b-401150059c-126215589 [Accessed 31/07/2018]

Logistics Manager (2018b) https://www.logisticsmanager.com/uk-to-remain-in-common-transit-convention/?utm_source=Subs_LM_Daily&utm_campaign=17e2470130-EMAIL_CAMPAIGN_2017_08_01_COPY_01&utm_medium=email&utm_term=0_27b19fa38b-17e2470130-126215589 [Accessed 21/12/2018]

Lowe, D and Pidgeon, C (2018) *Transport Manager's and Operator's Handbook 2018*, Kogan Page, London

Maersk (2018) https://www.maersk.com/global-presence/ [Accessed 29/11/2018]

Magaziner (2018) https://www.magaziner.de/fileadmin/magaziner/Dokumente/2018/03_Broschuere_EKBaureihe.pdf (in German) [Accessed 03/07/2018]

Mercedes (2018) https://www.mercedes-benz.co.uk/content/unitedkingdom/mpc/mpc_unitedkingdom_website/en/home_mpc/van.html?gclid=EAIaIQobChMIk7jmvK_-3AIVSbXtCh2EtgjxEAAYASAAEgLomfD_BwE&csref=web1606085001_ppc_100616&s_kwcid=AL!160!3!199275366135!e!!g!!mercedes%20vans&ef_id=WzHzXAAABeCQrxN_:20180821143739:s [Accessed 21/08/2018]

MK2 (2014) http://www.mk2.co.uk/property/peugeot-distribution-facility-tile-hill-coventry-cv4-9hs/ [Accessed 22/06/2018]

Monios, J and Bergqvist, R (2017) *Intermodal Freight Transport and Logistics*, CRC Press, Boca Raton, Florida. Non-print legal deposit e-book retrieved from www.bl.uk

Morrell, PS and Klein, T (2019) *Moving Boxes by Air: The economics of international air cargo*, 2nd edn, Routledge, Abingdon. Non-print legal deposit e-book retrieved from www.bl.uk

Motor Transport (2018a) www.motortransport.co.uk/blog/2018/03/23/gazely-gains-altitude-magna-park-takes-flight/ [Accessed 14/06/2018]

Motor Transport (2018b) https://motortransport.co.uk/wp-content/uploads/2018/01/MTR_111217_033.pdf [Accessed 28/08/2018]

MWPVL (2018) www.mwpvl.com/html/john_deere.html [Accessed 14/06/2018]

Narasimhan, G (2015) www.logisticssupplychain.blogspot.com/2013/09/outsourcing-1-to-10-pl.html [Accessed 12/06/2018]

Neise, R (2018) *Container Logistics: The role of the container in the supply chain*, Kogan Page, London

Network Rail (2018) https://www.networkrail.co.uk/the-positive-impact-of-rail-freight/ [Accessed 14/11/2018]

New Zealand Chamber (2018) https://www.newzealandchambers.co.nz/export-documentation/ [Accessed 11/12/2018]

New Zealand Government (2018) https://www.customs.govt.nz/globalassets/documents/tariff-documents/the-working-tariff-document-section-xvi.pdf [Accessed 19/12/2018]

Nissan (2018) https://www-europe.nissan-cdn.net/content/dam/Nissan/gb/brochures/Vehicles/Nissan_NV400_UK.pdf [Accessed 23/08/2018]

ONS (2018) https://www.ons.gov.uk/businessindustryandtrade/internationaltrade/articles/whodoestheuktradewith/ [Accessed 31/10/2018]

Pålsson, H (2018) *Packaging Logistics: Strategies to Reduce supply chain costs and the environmental impact of packaging*, Kogan Page, London

Passage East Ferry Company (2018) https://www.passageferry.ie/copy-of-buses [Accessed 27/11/2018]

PD Ports (2018) https://www.pdports.co.uk/documents/terms/pd%20teesport%20ukwa%20membership%202015.pdf [Accessed 18/12/2018]

Perry, C (2016) http://www.dailymail.co.uk/news/article-3515633/Why-women-refusing-lorry-drivers-forced-relieve-bush-says-transport-minister-blasts-haulage-firms-atrocious-toilets-truck-stops.html [Accessed 16/07/2018]

Pidgeon, C (2016) *A Study Guide for the Operator Certificate of Professional Competence in Road Freight*, Kogan Page, London. Non-print legal deposit e-book retrieved from www.bl.uk

Port of Dover (2017) https://www.doverport.co.uk/about/news/reefer-specialists-cool-carriers-bring-new-trade-t/13327/ [Accessed 19/09/2018]

Port of Felixstowe (2018) https://www.portoffelixstowe.co.uk/port/rail-services/ [Accessed 30/11/2018]

Port of Zeebrugge (2017) https://www.engie.com/wp-content/uploads/2017/06/regular_bunkering_op_pr_def.pdf [Accessed 03/12/2018]

QVC (2018) https://www.qvcuk.com/UK/images/vr_pdfs/QVC_A_Vendors_Guide_Final.pdf [Accessed 16/08/2018]

Rail Delivery Group (2018) https://www.raildeliverygroup.com/about-us/publications.html?task=file.download&id=469774087 [Accessed 14/11/2018]

Railway Age (2012) https://www.railwayage.com/mechanical/freight-cars/kasgro-builds-worlds-largest-railroad-car/ [Accessed 17/1/2019]

Railway Gazette (2001) https://www.railwaygazette.com/news/single-view/view/bhp-breaks-its-own-39heaviest-train39-record.html [Accessed 14/11/2018]

RAlpin (2018) http://www.ralpin.com/media/ [Accessed 14/11/2018]

Redirack (2018) www.redirack.co.uk [Accessed 22/06/2018]

Richards, G (2018) *Warehouse Management: A complete guide to increasing efficiency and minimizing costs in the modern warehouse*, 3rd edn, Kogan Page, London. Non-print legal deposit e-book retrieved from www.bl.uk

Richards, G and Grinsted, S (2016) *Logistics and Supply Chain Toolkit*, 2nd edn, Kogan Page, London. Non-print legal deposit e-book retrieved from www.bl.uk

Robins, N (2015) *The Ships That Came to Manchester*, Amberley Publishing, Stroud. Non-print legal deposit e-book retrieved from www.bl.uk

Sales, M (2017) *Air Cargo Management: Air Freight and the global supply chain*, Routledge, Abingdon

Selfridges (2016) https://images.selfridges.com/is/content/selfridges/Selfridges-Supplier-Guidelines-2016pdf/ [Accessed 16/08/2018]

Severn Boating (2018) http://www.severn-boating.co.uk/ryal.htm [Accessed 21/11/2018]

Shipbrokers' Register (2018) http://letter.wramfeltmaritime.se/salesform.aspx [Accessed 23/10/2018]

SMMT (2018a) https://www.smmt.co.uk/vehicle-data/lcv-registrations/ [Accessed 21/08/2018]

SMMT (2018b) https://www.smmt.co.uk/vehicle-data/car-registrations/ [Accessed 22/08/2018]

SMMT (2018c) https://www.smmt.co.uk/vehicle-data/heavy-goods-vehicle-registrations/ [Accessed 19/09/2018]

Software Advice (2018) https://www.softwareadvice.com/uk/scm/warehouse-management-system-comparison/p/all/ [Accessed 23/07/2018]

TfL (2018a) https://tfl.gov.uk/modes/driving/ultra-low-emission-zone [Accessed 18/10/2018]

TfL (2018b) https://tfl.gov.uk/info-for/deliveries-in-london/delivering-legally/accessible-roads-for-hgvs#on-this-page-0 [Accessed 01/11/2018]

Transit Owners' Club (2018) http://transitvanclub.co.uk/picture-history/index.html [Accessed 23/08/2018]

Travis Perkins (2018) https://www.travisperkinsplc.co.uk/~/media/Files/T/Travis-Perkins/documents/tp-plc-supplier-manual.pdf [Accessed 16/08/2018]

Trucking Info (2010) https://www.truckinginfo.com/150236/lift-axle-considerations [Accessed 18/09/2018]

Ubimax (2017) https://www.ubimax.com/cms/customer-projects-20170925-1-OXBrx9kW.pdf [Accessed 27/07/2018]

UK Government (2007) http://www.legislation.gov.uk/uksi/2007/871/contents/made [Accessed 01/08/2018]

UK Government (2015) http://www.legislation.gov.uk/uksi/2015/1640/pdfs/uksi_20151640_en.pdf [Accessed 01/08/2018]

UK Government (2017) https://www.gov.uk/government/news/2018-financial-levels-confirmed-for-commercial-vehicle-operators [Accessed 25/10/2018]

UK Government (2018a) www.gov.uk/health-conditions-and-driving [Accessed 29/11/2018]

UK Government (2018b) https://www.gov.uk/government/publications/notice-828-tariff-preferences-rules-of-origin-for-various-countries/notice-828-tariff-preferences-rules-of-origin-for-various-countries [Accessed 12/12/2018]

UK Government (2018c) https://www.gov.uk/government/publications/notice-104-ata-and-cpd-carnets/notice-104-ata-and-cpd-carnets [Accessed 14/12/2018]

UK Government (2018d) https://www.trade-tariff.service.gov.uk/trade-tariff/headings/8531?currency=EUR&day=18&month=12&year=2018 [Accessed 19/12/2018]

UK Government (2019) https://www.gov.uk/accepting-returns-and-giving-refunds [Accessed 07/01/2019]

UK Parliament (2015) https://publications.parliament.uk/pa/cm201617/cmselect/cmtrans/68/6805.htm [Accessed 12/10/2018]

UNCTAD (2015) https://unctad.org/en/PublicationsLibrary/itcdtsbmisc62rev6_en.pdf [Accessed 21/12/2018]

UNECE (2018) http://www.unece.org/tir/news/20181207.html [Accessed 21/12/2018]

Universal Cargo (2014) https://www.universalcargo.com/top-10-logistics-quotes/ [Accessed 08/01/2019]

Veolia (2019) https://www.veolia.co.uk/services/secure-destruction [Accessed 04/01/2019]

Volvo (2017) https://www.volvogroup.com/en-en/suppliers/our-supplier-requirements.html#KEP-document-download [Accessed 25/07/2018]

Volvo (2018) https://www.volvotrucks.co.uk/en-gb/home.html [Accessed 18/09/2018]

Wiegmans, B and Konings, R (eds) (2017) *Inland Waterway Transport: Challenges and prospects*, Routledge, Abingdon. Non-print legal deposit e-book retrieved from www.bl.uk

World Shipping Council (2015) http://www.worldshipping.org/industry-issues/safety/WSC_Summarizes_the_Basic_Elements_of_the_SOLAS_Container_Weight_Verification_Requirement___January_2015_-3-.pdf [Accessed 29/11/2018]

Note: In the UK, publishers are legally required to submit free copies of all books published to certain libraries, including the British Library. These deposits may now be made in electronic format, being viewable only in the reading rooms of the library. 'Non-print legal deposit e-book retrieved from www.bl.uk' indicates that books have been viewed in this way.

APPENDIX 1
Providers of products and services

A large number of products and services have been mentioned in this book, and I felt it would be helpful to list companies that are able to provide these. Appendix 1 includes websites of potential suppliers, mainly of organizations that have helped in the preparation of the book, for example by providing pictures (hence the inclusion of some without a direct connection to logistics). Please mention this book if you contact them. I have also listed a number of trade associations who may be able to assist, for example in identifying a suitable broker or forwarder.

Warehousing

Racking, shelving and mezzanine floors

Stakapal. Racking, shelving and associated safety products. https://www.stakapal.co.uk/

Gonvarri Material Handling. Racking, shelving, storage carousels, mezzanine floors, accessories. https://www.gonvarri-mh.com/

Avanta. Mezzanine floors, partitions, racking, shelving, accessories. www.avantauk.com

Storax. Racking, mezzanine floors. https://www.storax-group.com/en

Material handling equipment

Linde Material Handling. Wide range of forklifts, pallet trucks, pallet stackers, order pickers, tow trucks and automated trucks. https://www.linde-mh.com/en/

Crown Lift Trucks. New and used forklift trucks. https://www.crown.com/en-uk.html

Safety Lifting Gear. Pallet trucks, trolleys. Also a wide range of safety equipment, eg for working at height. https://www.safetyliftingear.com/

The Ramp People. Bridge plates and a wide range of ramps. https://www.theramppeople.co.uk/

Thorworld. Container ramps, loading platforms, dock levellers, accessories. https://www.thorworld.co.uk/

Loading Systems. Dock levellers, dock shelters, lifting platforms, doors, accessories. https://www.loading-systems.com/en-gb

Armo. Dock levellers, dock shelters, lifting platforms, doors, accessories. http://www.armo.uk.com/

Albion Handling. Conveyor systems. https://www.albionhandling.co.uk/

Axiom. Conveyor systems, packing systems, sorters. https://www.axiomgb.com/

B and B Attachments. Forklift truck attachments. https://www.bandbattachments.com/

Warehouse management systems

SAP. Full range of business systems, including WMS modules. https://www.sap.com/uk/index.html

SEP Logistik AG. WMS software and hardware. https://www.sepag.de/en/

Datalogic. Scanners, barcode readers, RFID systems. http://www.datalogic.com/

SSI Schäfer. Wide range of software and hardware, including WMS, scanners, RFID, semi-automated and automated picking. Also racking, mezzanines and conveyors. https://www.ssi-schaefer.com/en-gb

Barcode Warehouse. Barcode printers, scanners and software. https://www.thebarcodewarehouse.co.uk/

Miscellaneous

Universeal. Container seals and other security seals. https://www.universeal.co.uk/

Flogas. Bulk and cylinder gas for forklift trucks. https://www.flogas.co.uk/

Herchenbach. Temporary warehouses. https://www.herchenbach.co.uk/

Packaging and related equipment

Macfarlane Packaging. Wide range of packaging and voidfill products, and associated machinery. https://www.macfarlanepackaging.com/

Robopac. Stretch wrapping, shrink wrapping, taping and case erector machines. http://www.robopac.com/UK/

Rajapack. Range of packaging materials and machinery. https://www.rajapack.co.uk/

Yolli. Food packaging, and packaging equipment. https://www.yolli.com/

Davpack. Packaging materials and equipment. https://www.davpack.co.uk/

Barnes and Woodhouse. Wooden crates, pallets and boxes. https://www.timberpackingcases.com/

Premier Pallet Systems. Pallet inverters and dispensers. https://premierpalletinverter.co.uk/

Lowe RP. Metal stillages and cages. https://www.stillagesandcages.co.uk/

PPS Equipment. Returnable packaging, including one-way hire and washing. https://www.ppsequipment.co.uk/

Vehicles

Volvo. Rigids, chassis cabs and artics, 10 tonnes and above. https://www.volvotrucks.co.uk/en-gb/home.html

Isuzu. Pick-ups. https://www.isuzu.co.uk/ Rigids and chassis cabs 3.5 to 13.5 tonnes. https://www.isuzutruck.co.uk/

Ford. Vans, chassis cabs and pick-ups, 1.5 tonnes and above. https://www.ford.co.uk/vans-and-pickups

Scania. Rigids, chassis cabs and artics, mainly 18 tonnes and above. https://www.scania.com/uk/en/home.html

Sotrex. Used vehicles. https://www.sotrex.com/

LC Vehicle Hire. https://lcvehiclehire.com/

Dennison Trailers. Specialist in skeletal and platform (flatbed) trailers. http://www.dennisontrailers.com/

Vehicle-related products and services

BigChange Advisory Plus. Transport and workforce management software. https://www.bigchangeapps.com/

Maxoptra. Vehicle routing software. https://maxoptra.com/

TruTac. Tachograph analysis software. https://www.trutac.co.uk/

Brendeck. Vehicle wash systems. https://www.brendeck.co.uk/

Checkpoint Safety. Wheel nut indicators. https://checkpoint-safety.com/

UK Drug Testing. Workplace drug and alcohol testing kits. https://www.uk-drugtesting.co.uk/

Roadfreight

Hallett Silbermann. Heavy haulage and abnormal loads specialist. http://hallettsilbermann.com/

Eddie Stobart. Wide range of road, rail and warehousing services. http://eddiestobart.com/

Annandale Transport. Forestry specialists. https://www.annandaletransport.co.uk/

Denby Transport. Range of transport services. http://www.denbytransport.co.uk/

Palletline. UK and European pallet delivery. https://www.palletline.co.uk/

Abbey Logistics. Bulk tanker specialist. https://www.abbeylogisticsgroup.com/

Goldstar. Container haulage specialist. https://www.goldstartransport.co.uk/

Gefco. Range of UK and international transport and logistics services. https://uk.gefco.net/

Stears Haulage. Flatbed trailer specialists. https://stearshaulageltd.com/

Avon Material Supplies. Tippers and grab lorries in Southern England; supply of aggregates and building materials. https://www.avonmaterialsupplies.co.uk/AMS/

Railfreight and inland waterways

Freightliner. Intermodal and bulk haulage. https://www.freightliner.co.uk/

Network Rail. Operator of UK rail infrastructure: tracks, signalling, etc. https://www.networkrail.co.uk/

Transmec. Intermodal services – also road, sea and air. https://www.transmecgroup.com/en/home

Eurotunnel. Operator of the Channel Tunnel Le Shuttle services UK to France. https://www.eurotunnel.com/uk/home/

RAlpin. Rolling highway services across the Alps. http://www.ralpin.com/

Mercurius. European inland waterway shipping. https://www.mercurius-group.nl/ (website is in Dutch)

Seafreight and airfreight

CLdN. Bulk and ro-ro shipping. http://www.cldn.com/

Unifeeder. Short sea and feeder services. https://www.unifeeder.com/corporate

Scotline. Short sea liner services and vessel charter. http://www.scotline.co.uk/

Port of Zeebrugge. Full range of port services. https://www.portofzeebrugge.be/en

Chapman Freeborn. Global aircraft charter specialists. https://www.chapman-freeborn.com/en/about-us/

Other products and services

Motor Transport. Periodical serving the roadfreight industry. https://motortransport.co.uk/

Marlow, Gardner and Cooke. Insurance brokers. https://www.marlowgardner.co.uk/

Techsert. Export certification for former Soviet block. https://www.gostrussia.com/en/

Prologis. Developer and owner of industrial and logistics buildings. https://www.prologis.co.uk/

Abbott Wade. Staircase installations. https://www.abbottwade.co.uk/

Pets at Home. Pet supplies retailer. http://www.petsathome.com/

Chambers of commerce

International Chamber of Commerce (publisher of Incoterms®). https://iccwbo.org/

Eurochambres. The Association of European Chambers of Commerce and Industry – website lists chambers of commerce throughout Europe, excluding the UK. http://www.eurochambres.eu/Content/Default.asp?pagename=OurMembers

Arab Chambers. Website gives details of chambers in Arab countries. http://uac-org.org/en/ArabChamber/arab-chambers

UK. https://www.britishchambers.org.uk/

USA. https://www.uschamber.com/

Australia. https://www.australianchamber.com.au/

New Zealand. https://www.newzealandchambers.co.nz/

Canada. http://www.chamber.ca/

China. All China Federation of Industry and Commerce. http://www.chinachamber.org.cn/

Others. Links to other Chambers of Commerce can be found via this website. https://www.chamber-commerce.net/

Trade associations

Freight forwarders

BIFA. UK. https://www.bifa.org/home

FENEX. Netherlands. https://www.fenex.nl/ (in Dutch)

AICBA. Canada. https://www.aicba.org/

CIFA. China. http://www.cifa.org.cn/cifa_En/

HAFFA. Hong Kong. http://www.haffa.com.hk/portal/

JAFA. Japan (air cargo). http://www.jafa.or.jp/english/

NAFL. Dubai. http://nafl.ae/

Others

International Air Transport Association. www.iata.org

Institute of Chartered Shipbrokers. https://www.ics.org.uk/

Freight Transport Association. https://fta.co.uk/

Road Haulage Association. https://www.rha.uk.net/

United Kingdom Warehousing Association. https://www.ukwa.org.uk/

Storage Equipment Manufacturers Association. https://www.sema.org.uk/

APPENDIX 2

Brexit

I should say something about the exit of the UK from the European Union. I had expected that by the time I finished writing this book (late January 2019) there would be a clear picture of what form Brexit would take. Unfortunately, we are less than two months from the scheduled date for Brexit, and we are none the wiser: we cannot be sure about what will happen on that date, or at any time thereafter, as the UK moves from a transition period to more permanent arrangements. There have been predictions of shortages of food and even medicines, and of massive queues of vehicles around Dover. We do not know what customs formalities will be required at the border. However, if these formalities prove to be complex and/or time consuming, there will be insufficient staff to carry them out, and they will not be able to be recruited and trained at minimal notice. The question of the land border in Ireland is also unresolved.

I would love to be able to give meaningful advice on the impact of Brexit. Sadly, however, neither I nor anyone else can do so at the present time.

INDEX

Italics indicate figures or tables.